The Shaping of Life

Biological development, how organisms acquire their form, is one of the great frontiers in science. While a vast knowledge of the molecules involved in development has been gained in recent decades, big questions remain on the molecular organization and physics that shape cells, tissues and organisms. Physical scientists and biologists traditionally have very different backgrounds and perspectives, yet some of the fundamental questions in developmental biology will only be answered by combining expertise from a range of disciplines. This book is a personal account of an interdisciplinary approach to studying biological pattern formation. It articulates the power of studying dynamics in development: that, to understand how an organism is made, we must know not only the structure of its molecules; we must also understand how they interact and how fast they do so.

LIONEL G. HARRISON (1929–2008) was Professor Emeritus in the Department of Chemistry, University of British Columbia, where he was a Faculty member for 50 years. A physical chemist by training, he was inspired to study biological form and 'in developmental biology . . . found something different and immensely exciting: a field with a Great Unknown' as he wrote in his 1993 book, *Kinetic Theory of Living Pattern*. 'To pursue it is like trying to account for the rainbow in the 14th century, to do celestial mechanics before Newton, or to pursue quantum theory in the 1890s.'

The Shaping
of Life

The Generation of
Biological Pattern

LIONEL G. HARRISON
University of British Columbia
(1929–2008)

CAMBRIDGE UNIVERSITY PRESS

Cambridge, New York, Melbourne, Madrid, Cape Town, Singapore,
São Paulo, Delhi, Dubai, Tokyo, Mexico City

Cambridge University Press
The Edinburgh Building, Cambridge CB2 8RU, UK

Published in the United States of America by
Cambridge University Press, New York

www.cambridge.org
Information on this title: www.cambridge.org/9780521553506

First published 2011

Printed in the United Kingdom at the University Press, Cambridge

A catalogue record for this publication is available from the British Library

Library of Congress Cataloging-in-Publication Data

Harrison, Lionel G.
    The shaping of life : the generation of biological pattern / Lionel G. Harrison.
        p.   cm.
    ISBN 978-0-521-55350-6 (Hardback)
    1. Pattern formation (Biology)   2. Morphogenesis.   3. Developmental
biology.   I. Title.
    QH491.H35 2011
    571.3–dc22

                                                                    2010029820

ISBN 978-0-521-55350-6 Hardback

Contents

Foreword

The Shaping of Life is Lionel's second book, completed shortly before his death in March 2008. It is in part an update of his earlier volume, *Kinetic Theory of Living Pattern* (Cambridge University Press, 1993), following developments in the field and in his own research. Lionel was always an energetic individual, both physically and intellectually, climbing mountains and arguing points of scientific theory with equal determination. Those interested will find a tribute covering salient aspects of his personal and professional life in *The Globe & Mail* for 26 April 2008. He experienced a great loss in the early 1980s with the deaths, within the space of a few weeks, of both his wife and only son. His life afterward was increasingly centred on scientific activities, and he continued with research to retirement and after. This book is a chronicle of his observations and insights during that time, and of the people who influenced him and helped along the way. His proximate goal was to discover how patterns of living things are formed or, to use the title from an earlier draft of this book, 'how life devises its shapes and sizes'. Though it contains considerable experimental detail, the book is addressed to a non-specialist audience, and especially to those interested in how science is done in a field still in embryo, whose mature form is as yet unknown.

Working in an emerging area of science is a challenge and an adventure, and can be terribly exciting. There are, however, definite risks to reputation and career, as the methods may be unconventional or untried, and there is often no clear criterion for what constitutes progress. Cautious academics will avoid such fields until the ground rules are better worked out, and until it is clear that a predictable and, hence, grant-worthy rate of progress can be sustained. Coming from the relatively mature discipline of surface chemistry, Lionel, by contrast, was immensely excited by the prospect of doing research on aspects of biology still at a very early stage of their development. It was, to him, like doing 'celestial mechanics before Newton'. All things were possible, and even very simple observations might lead to far-reaching insights. Like others in the physical sciences, he was particularly struck by the elephant in the room, one that working biologists are usually loath to acknowledge: that biology today, excepting the great insight of Darwin, is essentially a descriptive science devoid of a set of fully developed principles. Most physical scientists assume this situation is simply temporary, and that biology will proceed along the path already taken by physics and chemistry, from the specific and descriptive to the synthetic and analytical, using, eventually, a mathematical formulation.

What will this 'new biology' look like? First, it will need to improve on the vague way we currently deal with the concepts that, so far as we can tell, must lie at the core of biology. 'Organization' is one example. 'Order' is another, including spatial pattern, but also the ordering of genomic and metabolic networks in living cells, as is 'differentiation', the process by which such things change over time. None of these is as yet more than a name attached to a notion, and none are properly defined and measurable. What, for example, are the dimensions (units of measure) of organization, and what precisely is its role in a subsidiary process like pattern formation? To answer this, one needs to be quantitative, so that the amount of the former (organization) that is required for, involved in or used up by the production of the latter (pattern) can be calculated. Only then do the words take on concrete meaning, and only then can the analysis be fruitfully carried forward. Consider, for example, the production of the five-digit human hand from its flattened, paddle-like beginnings: this is generally accepted as a form of differentiation, and the end product does look more ordered, or at least anatomically more complicated, than the beginning. But the cells involved, of cartilage, connective tissue and skin, are still doing pretty much the same jobs, only the locations have changed. So how much ordering has actually occurred, and if this is truly differentiation, is it a trivial amount in quantitative terms, or something worth accounting for in relation to the cost paid in disorder by the rest of the universe? Thermodynamics is the way to deal with such questions, but we as yet have only the most rudimentary beginnings of a thermodynamics applicable to the complexities of living systems. Lacking this, Lionel's approach, best exemplified in his *Acetabularia* experiments (Chapter 3), was necessarily empirical, and as such, entirely characteristic for a physical chemist: measure patterns and pattern change directly, vary the conditions over which you have control, and analyse the outcome.

Here he came closest to putting meaningful numbers on an otherwise mysterious process, deducing such exotic thermodynamic beasties as ΔH and ΔS from measurements requiring only a dissecting microscope, thermometer and micrometer measuring scale. A complete thermodynamics of living systems is clearly some distance in the future, but of various routes to that future, this is certainly one. Lionel's book is, in this respect, as much an object lesson in scientific methodology as an exploration of pattern in nature.

Lionel's wide-ranging interests within and outside science are evident in his writing, which is liberally supplied with references to literature, historical anecdote and, especially, music. The operas of Wagner were a particular interest, especially where these touched on issues of creativity and intellectual challenge. Two of his favourites: first, Act I of *Die Meistersinger*, which beautifully captures the unavoidable and sometimes comic conflict inherent in any human endeavour, but especially academic ones. On the one side is accepted convention and governance by rules, here represented in the character of Beckmesser. On the other, the need for periodic inspiration and renewal through the intervention of a more gifted practitioner – the headstrong young knight Walther von Stolzing in this case. No prize is offered for guessing how Lionel saw himself in this context. A second favourite was Act I of *Siegfried*, which offers a different sort of confrontation, between Mime, a dwarf, and Wotan, King of the Gods, here in the guise of the one-eyed Wanderer. Mime is obsessed with the task of re-forging the magical sword *Notung*, which, despite his mastery of dwarvish technology, he is unable to do. Wotan knows precisely what is needed, but given the chance to question Wotan, Mime instead wastes the opportunity on set questions for which he already knows the answers. For scientists confronting Nature, and especially those outside the confines of an established discipline, one needs bold questions indeed. Lionel's questions were always as bold as he could make them, and he clearly hoped by example to encourage others to do the same.

Thurston Lacalli
Victoria, May 2009

Acknowledgements

Lionel inspired a great many of us over the years with his energy and brilliance. It is with great pleasure that we are able to present the fruits of his last decade of labour, and hopefully inspire many more with his ideas and questions.

It took the efforts of many friends and colleagues to bring this book to completion. Thurston Lacalli (in addition to writing the Foreword), Michael Lyons and Harold Kasinsky made careful readings and many comments on the text, as Lionel left it in February 2008. Axel Hunding supplied some of the updated references in Chapters 8 and 9. Members of the UBC Chemistry Department have been extremely helpful and supportive – to single out a few, but not exclude the many others: Elliott Burnell, Ed Grant, Elena Polishchuk, Nick Burlinson, Gren Patey, David Walker, Yoshi Koga and Elizabeth Varty, who helped on many of the figures. Thanks go to the Belkin Gallery at UBC and the Shadbolt estate for the cover image.

I have coordinated this job, finalized the text and figures and written the Epilogue to provide a current context. I have done minimal editing within the text; rather opting to insert comments and bring references up to date in footnotes throughout the book. The aim is to let Lionel's ideas shine through, and give pointers to further reading in the current literature. I am very grateful for the support and guidance of the Cambridge University Press production team, in particular Katrina Halliday and Lynette Talbot. My institution, the BC Institute of Technology, has supported my time on this work. The outpouring of support and remembrances by Lionel's friends and colleagues is a testament to his life; and it has made finishing this book much easier. To Lionel.

David Holloway
Burnaby, April 2010   xi

Having used 'once upon a time' to start the preface of my previous Cambridge University Press book, *Kinetic Theory of Living Pattern* (1993), what else could I do but use a Lewis Carroll quotation as the epigraph to the preface of its sequel,[1] especially since my interest in mechanisms of biological development began from the looking-glass problem of optical resolution? As a physical chemist, hitherto concerned only with inorganic materials, I was certainly wandering into unknown territory, demanding of me a grand survey that I was certainly not about to make when, through 1972–3, a couple of chance occurrences redirected me into this field of enquiry. These, as mentioned in my 1993 preface, were attending a talk by my colleague R. E. Pincock on an instance of 'spontaneous optical resolution' that he had observed in careful experiments, and my first introduction to studies of biological morphogenesis by taking the chair at the PhD defence of Thurston Lacalli, who received a degree in zoology by working on development of an alga. Those two afternoons led me to get excited about the concept of 'non-linear dynamics' as probably being a major part of the explanation of both classes of phenomena. I might quite easily have decided that I could attend neither of these events; and in that case, I might never have undertaken any work in this field. My first enthusiastic thanks must therefore be to both of these gentlemen, and to whosoever among the gods *does* play dice in arranging human encounters. (For further commentary on optical resolution, non-linear dynamics and gods playing dice, see the introductory paragraphs to Part II.)

From the later 1970s, I was greatly encouraged by the late Paul Green, a botanist who insisted that 'there is no escaping the calculus when studying development'. He proposed the writing of my previous book, and is hence responsible (at one remove) for the existence of this sequel. His death from a rapidly fatal cancer in 1998 was the loss to biology of an important scientist whose work had not yet reached its proper fulfilment, and to me also of a very congenial friend.

In the projects from my own work that I describe here, much was collaborative and I have many thanks to give to: Beverley R. Green, University of British Columbia, a biochemically trained botanist whose work on the alga *Acetabularia* gave me a highroad into experimental biology; John B. Armstrong, University of Ottawa, who studied

---

[1] These acknowledgements of Dr Harrison's are a year older than his final version of the text; the final preface did not have the Carroll quote. Taken from *Through the Looking Glass*, it is: 'Of course the first thing to do was to make a grand survey of the country she was going to travel through.'

embryonic development of the axolotl (a salamander) and got sufficiently interested in the possibilities of reaction–diffusion theory in
this topic that the exploration of that possibility made most of the
PhD thesis of my student, David M. Holloway, who remains a close
collaborator; Patrick von Aderkas, University of Victoria, BC, who
worked experimentally on somatic embryos of conifers, giving me the
chance to do some measurements myself and find the Bessel functions
in the seed leaves; Harold E. Kasinsky, University of British Columbia,
who kept bothering me about patterns of condensing DNA in sperm
until I got to believe that even these could have a dynamic explanation,
and collaborated with him and his collaborators Manel Chiva and Enric
Ribes at the University of Barcelona; Jacques Dumais, now at Harvard,
whose Master's thesis work with me led on to a PhD with Paul B. Green
at Stanford University and continuing interaction on mechanical forces
in plant surfaces; and once again but fully deserving a second mention,
Thurston C. Lacalli, successively of the Universities of British Columbia,
Saskatchewan and Victoria, whose intensive collaboration with me for
several years in the 1970s, which has continued less intensively ever
since, was the way we both developed our understandings in some
depth of Turing's 'chemical basis of morphogenesis'.

Lionel Harrison
Vancouver, November 2006

Preface

This book is intended for anyone who is interested in contemplating the question posed in the title,[1] and who has a modicum of general scientific education. It is not directed specifically to people working in the sciences; and of those who are, it is not specifically for physical or biological scientists. It is mainly for people who would like to think about unsolved questions rather than to receive answers. Thus, it is not a review of any specific specialized field, and particularly not of those aspects of biology that have recently been producing answers most rapidly – the aspects to do with genes and genomes and the daily increasing number of words ending in the sacred syllable –ome. My subtitle, 'the kinetic aspect',[2] implies that I am interested much more in how rapidly things happen to various objects than in what the objects are. Further, I believe that the study of rates of change can often be pursued as a primary objective, independent of knowing in advance the material composition of the objects that are changing. (Perhaps not only the primary objective, but also the ultimate – philosophers have long pondered whether the deepest understanding of the universe must be in terms of matter or motion.) This implies that if I consider a biological phenomenon and have the urge to find out 'what is doing it', I am not usually trying to find the name or formula of a substance or molecule, but rather the forms of some expressions showing how the amounts or concentrations of a few substances must be changing in space and time.

This attitude can, however, seem old-fashioned, and I am vulnerable, like all emeritus professors, to the sneer that I have passed the

---

[1] The question in an earlier draft was 'How does life devise its shapes and sizes?'

[2] This was an earlier subtitle, referring back to Dr Harrison's 1993 book, *Kinetic Theory of Living Pattern*.

'philosopause' (a word I learned as current from Natalie Angier's *The Canon*, 2007). Indeed, my attitude can be seen as representing the way D'Arcy Thompson expected biology to advance when he first wrote *On Growth and Form* in 1917 (the frequently referenced version of 1942 was the second edition). Neither he nor anyone else at that time anticipated that in the twentieth century, via electron microscopy and molecular chemistry, anatomy would advance to its ultimate limit of minuscule size, the very sequences of molecules and hence of atoms in living beings. I think also that many scientists intensely concerned with continuing the elaboration of such detailed descriptions, and therefore immersed in the work of the past few years or even months, may find something apparently philosopausal in the habit of theoreticians to reach back to treatments of kinetics or diffusion or vibrations of discs or whatever was published between the 1920s and 1950s. But this is a necessary feature of such work; to be sure, we do not regularly reference Newton's original writings of three and a half centuries ago whenever we relate forces to accelerations, but 60 years or so ago is yesterday. When theory has been founded upon rocks, we have to keep on digging back all the way to the rocks as the foundations on which to build.

It has been opined to me by a reviewer of the penultimate draft of this book that biologists reading it 'are going to be hostile because of the fact that genomes and bioinformatics etc do not get much of a treatment'. I can see that many people absorbed in those popular branches of biology might be uninterested in my topic, and might therefore decide not to read this book beyond this sentence. But I do not understand hostility to it, because my topic is entirely complementary to the molecular aspects, and in no way inimical to them. (Nevertheless, I have indeed encountered negative attitudes, as indicated at places in my account.) Still less do I comprehend the attitude well described by the same reviewer in the words, 'the biological world, almost hysterically, embraces "systems biology" with a zeal and faith that can only be described as "religious"'. The attitude thus defined disturbs me in two ways: first, in my home discipline of chemistry, people working in manners ranging from almost a mathematician's to almost a biologist's and everything in between are not only tolerated but welcomed into the community of the discipline. I appreciate the chasms across which different people gaze at each others' viewpoints, often without finding the means to cross over, but not why this should lead to hostility. The source of enmity is more usually covetousness for land that is easy for both opponents to grab because there is no chasm to cross. But second, even in regard to the analogue of religion, I hold pantheistic views that, when two religions are such that it seems necessary for the adherents of one to regard the other as necessarily untrue, I am inclined to think that both have equal quanta of the truth, but neither has the monopoly on it that its adherents think.

Let's get back to the chasm. There is one, between dynamical theorists and molecular biologists, and it is desirable that it should soon be spanned by one or more bridges in other than ramshackle shape, so that equations of motion and structural details of genes and proteins can become parts of the same territory in which everyone is free to wander. Living organisms have for centuries (or even millennia) provided much inspiration for research in all of the sciences (even astronomy, when its practitioners go looking for distant planets likely to have life-supporting conditions or little shreds of organic stuff littering the solar system).

Many of these sub-disciplines need no mutual interaction. But for finding out how organisms organize their development, a junction between molecular detail and the vastly larger macroscopic scale of spatial patterning does need to be built. It has been suggested to me that I should write a chapter describing that bridge. I both appreciated the suggestion and found it quite alarming, because it amounted to describing in some detail the middle of something of which only the extreme end sections have yet been built by anybody. But one should never refuse the challenge to write an essay about nothing. And while bridges are being built, there are structures in place at crucial stages called the falsework. They are made of wood, and show where the bridge is going to be, but give very little idea of what it is eventually going to look like, and no idea at all of what it will be made of. I can describe the existing falsework adequately in this prefatory account; and quite recently, at the time of writing, it has acquired at least one thin line that seems to span the entire chasm.

To begin at my end: dynamics can be studied for instances involving known substances, sometimes even for reactions with known values of rate constants. In the general realm of cell biology, it is becoming fairly popular for applied mathematicians to tackle such systems. A contrasting approach is to characterize the properties of particular kinds of dynamics without regard for the particular matter that may display them. The simplest well-known example is what is variously known as an exponential decay, first-order decay or relaxation process. It has mathematical properties, such as the constant time for a concentration to halve, that are the same whether one is thinking of radioactive decay, degradation of a protein or free induction decay of a set of nuclear spins in an NMR experiment. It is important to many types of scientist to know the properties of such dynamics.

I contend that, for the phenomena of developmental biology, the scope of pattern-forming dynamics should likewise be common knowledge, at least in fairly general terms. The route towards this knowledge has been well indicated by Hans Meinhardt, especially in his book on sea-shell pigmentation patterns (1995, 1998, 2003 editions), which makes some characteristics of solutions of partial differential equations accessible at the level of coffee-table presentation. I have seen Meinhardt's approach described (by an anonymous reviewer) as 'toy models', a term used by some trendy physicists to downgrade what I think of as a proper route into theory by setting out hypotheses.

That approach stems from four essential preconceptions (or paradigms, for any who prefer that much-misused word):

(1) a pattern, as formed by an event in biological development, frequently starts as a single entity on the macroscopic scale, often tens of micrometres between repeating parts in examples as diverse as whorls in single-celled algae (Chapter 3) and somites in vertebrate embryos, including human (Chapter 9). The pattern often starts out as a harmonic spatial waveform (Sections 3.2 and 4.4), and needs to be studied as an entity.

(2) These events are capable of setting up quantitative scales of distance, and therefore, wherever possible, it is useful to characterize the events by spatially quantitative measurements. This is good for a physical chemist blundering into biology and wanting to do experiments, because much can be done by high-school methods at low magnifications. One graduate student I converted from a chemist to a developmental biologist remarked that I had taught him that research can be done with a dissecting microscope.

(3) The mechanisms that control such pattern formation are likely to be matters that it is the proper business of physics and physical chemistry to address, with, therefore, a modicum of mathematical language used in even the first tentative interpretation of the experimental results.

(4) Many patterns may be kinetically generated and kinetically maintained, and theories of their formation must deal in depth with rates of reaction and transport of substances

in mechanisms belonging to non-linear dynamics (and mechanical forces may also be involved in ways that demand mathematically expressed theory).

The obvious basis for the start (in 1973) of my own work in this field was the well-known (but still to my mind lamentably under-used in developmental biology) theory of pattern generation by activator–inhibitor interactions initiated by Alan Turing (1952) and elaborated by Prigogine, Meinhardt and others so that 'reaction–diffusion' is now a familiar and extensively studied branch of applied mathematics. Should this type of theory, in which no more than a handful of substances do the whole job of making complex repeated patterns out of uniformity, be thought of as out-of-date when enormous networks of genetic interactions now loom over the discipline and it is repeatedly said that no-one has positively identified a Turing morphogen pair? The latter point brings me back to my example of a thin line that now seems to span the chasm: a group in Freiburg (Sick *et al.* 2006) have made the definite statement that 'WNT and DKK determine hair follicle spacing by a reaction–diffusion mechanism'. That paper identifies these two proteins as activator and inhibitor in a Turing mechanism (see Section 6.3.5).[3]

As to the question of a few substances versus a vast network, I believe that knowledge of what a few substances can do dynamically with a particular pattern of interactions between them is akin to having a divining rod that, on a wander through the tangles of the network, will twitch when one reaches the small region that has what it takes to make spatial pattern. Here, perhaps, is some reason for the 'hostility' that I have mentioned as puzzling me. The twitch may delight the theorist anxious for a source of equations that quench the thirst for unification and understanding, but cast into despondency the careful cartographers of a broad landscape of genes and proteins, who are being told that most of their country has only dry wells for the particular pattern-forming phenomenon being considered. Fear not, cartographer, nor cast out the theorist into the unmapped desert. The gene you have discovered will turn out to be nectar in another pattern-forming event that neither of you has yet ever thought of.

But what do I think is necessary to make a divining rod that twitches at the right places? Does it have to have a phoenix feather

---

[3] See also Digiuni *et al.* (2008), identifying GLABRA and TRIPTYCHON as acting through Turing dynamics to control the spacing of trichome structures on leaf surfaces.

embedded in it, or anything that the experimental biologist will believe equally mythical and unlikely to be findable? In other words, what different kinds of experiments need to be done? To approach this kind of question, I was effectively forced into making large parts of this book effectively a memoir of large parts of the work of me and my group since about 1973 (most of Part I, and Section 9.1). I have already stated (as my 'essential preoccupation (2)' above) that much of this involves spatially quantitative work on a large enough spatial scale that it requires only low magnifications and may seem to be a reversion to high-school biology. The essence of this kind of work is that it is like classical chemical kinetics (yes, of the 1930s). One does not leave a system to make something and come back later to see what it has made. One keeps on observing and measuring as a function of time, and particularly in the hope of catching the system at exactly the point at which it seems to be making the rudiment of the pattern. Geneticists tend to avoid any encouragement I give them to look for mutants in pattern formation, because what I want to see is no spectacular change in development, but, for instance, a quantitative change in the spacing between repeated parts in a pattern; and that is in no way easy to spot, except by multitudinous painstaking measurements, most of which may turn out to be wasted effort.

From the first encouragement I received from Cambridge University Press editors to write this book, it was intended to be in a more 'popular' style than my 1993 book – 'popular' meaning readily readable by anyone with fairly substantial education in *any* of the sciences. In the earlier book, I tried to hit a level of mathematical presentation accessible to anyone with a good first course in calculus, which wasn't difficult since I have never taken a university course in mathematics and have always been coasting along on a knowledge of calculus from sixth-form high school in England in 1944–6. But even that level of presentation led me to put 143 numbered equations in Chapters 6, 7 and 9 of that book. (One does not need a high level of sophistication to be unconscionably verbose, or should I say 'equationose'.)

The present book contains only a handful of equations. (Despite having heard that Stephen Hawking was told, when he was writing *A Brief History of Time*, that putting in even one equation halves the sales of the book, I couldn't get rid of them all.)

This account is, above all, interdisciplinary; I advocate attempts to bring together some aspects of the physical and biological sciences for the common aim of finding out by what means biological organisms go about developing the myriad complexities of their shapes. And one only brings together aspects by bringing together people. What difficulties does this involve? First, there is language. Science, as a whole, has become enormous, and therefore divided into numerous almost completely separated specialties, each with its own little language unknown to any of the others. Twenty-first-century science has far surpassed the Tower of Babel in this property. Science is becoming a multicellular organism with practitioners like molecules that can't cross membranes. I am disturbed at the number of seminars I attend at which the speakers know they are addressing general audiences but have made no attempt to adjust their use of specialized terminology to that situation. A lot of time has to be devoted to this if the Tower of Babel is to become syncytial, with passages like gap junctions or plasmodesmata between all its rooms. (Now, how many readers have I lost with the terminology in that sentence? I'm just talking about routes by which substances manage to pass from cell to cell.)

Never let it be said that the chasm between the sciences is just that biologists can't do mathematics. The barrier is both ways across every interdisciplinary boundary. But I still believe that many more developmental biologists must make the effort to understand somewhat more mathematics if some essential aspects of their discipline are to advance at all. As I understand the meaning of 'science', the essence of it lies in putting experiment and theory together. Far fewer people can achieve real science than can do reliable experiments or manipulate mathematical formalisms well; the greater part of science lies in putting these two together. In physics and chemistry, that union can commonly be made at a social level, with any particular individual identified as an experimentalist or a theoretician. Not so, I believe, in developmental biology at the present time. The connection between theory and experiment is still at a rudimentary level at which totipotency in the scientific enterprise needs to be present at the level of the small research group. While I advocate more knowledge by everyone of what's on the other side of the fence, I don't generally anticipate that plant-breeding in culture vessels and model-breeding in computers will be done most usually by the same person. The interdisciplinary perspective needs only to be enough that both people can talk to each other to the extent of designing research projects together, and be trained enough in each others' disciplines to understand their relative preconceptions and expectations and be interested in talking!

With the above list, serving as acknowledgement and as advocacy of small-group but broad-perspective collaborations, I may unwittingly have provided evidence to condemn me as an incurable dilettante, who never stays long with work on one organism. I prefer to say that I thoroughly deplore the concept of basing biology chiefly on the study of a few 'model organisms'; that one cannot establish the unifying nature of concepts in biology other than by looking for them in a wide range of biological diversity; and that the value of the comparative approach was best put a long time ago by William Harvey (1578–1657), who established the concept that blood circulates, and wrote:

> The common practice of anatomists in dogmatizing on the general make-up of the animal body, from the dissection of dead human subjects alone, is objectionable. It is like devising a general system of politics, from the study of a single state, or pretending to know all agriculture from an examination of a single field. It is fallacious to attempt to draw general conclusions from one particular proposition . . . . Had anatomists only been as conversant with the dissection of the lower animals as they are with that of the human body, the matters that have hitherto kept them in a perplexity of doubt would, in my opinion, have met them freed from every kind of difficulty. (copied as quoted by Crombie 1953, from Chapter 6 of Harvey's *De Motu Cordis, On the Movement of the Heart*).

Following Harvey's attitude perhaps even more broadly, I am quite happy to present an account of developmental concepts, the greater part of which is about plants when the greater part of biology today is about animals. Where would our knowledge of genetic diseases have got to without Mendel's groundwork on peas? In regard to the concepts of developmental theory that I like to pursue, a big advantage of plants is that, because of the rigidity of their cell walls, they are more amenable than animals to quantitative spatial measurements on patterns as they form. Paradoxically, this implies that organisms that display the least cellular motion within tissues during development may be the best for studies seeking to explore the roles of dynamics in pattern formation.

Since I wrote my 1993 book, there have been some substantial advances in the recognition that there are places for mathematical work in biology. University departments are establishing mathematical biology groups and making faculty appointments in them. The complete determination of genome composition in a number of species has led, on the one hand, to another version of stamp-collecting in the matter of proteomes, but also to the recognition that when one has all the structures, the next focus of attention should be on the functions of all these molecules. And that means looking at changes, and hence rates of change, and hence using differential equations. Or am I over-interpreting the trends with a bit of wishful thinking? I tend to get interested when

I hear (increasingly often) the terms 'genetic networks' and 'systems biology', and then to be rather disappointed when I find that what is being presented is not approaching my interests at all directly.

But surely, schematics of networks of interactions between genes via, for instance, a protein product of one gene activity being a transcriptional regulator for the activity of another gene, are likely to contain many of the processes having non-linear dynamics capable of pattern formation. (I write 'many', not 'all' of the processes because I believe that pattern formation may often involve biochemistry very distant from such relatively direct interactions of genes; see my opinions on identities of Turing morphogens in Section 7.3.3.) Between thousands of genes there are potentially millions of pairwise interactions. The kinds of dynamics my work is devoted to can produce complex spatial patterning out of the four self and mutual interactions of two substances. To find these in a network can be the needle-in-a-haystack problem; but this needle must be found because a prick from it can change the shape of the whole haystack.

To clarify this point, in Section 1.5, especially Fig. 1.1, I compare two diagrams: one is of my devising (Harrison 1993, Fig. 1.1). It has four columns; the second column from the left contains a partial differential equation. The other figure is from Bornholdt (2005), in an article entitled 'Less is more in modeling large genetic networks'. It has four columns; the second from the left contains two ordinary differential equations. There is at some level a similarity between the approaches symbolized in these two schematics, and at other levels strong contrasts. I present this, and everything else in the following pages, to let readers make up their own minds what approaches are going to lead to bridge-building between people from diverse academic disciplines all fascinated by the shaping of life. One of the commonest and simplest-looking events around us is the branching of one plant stalk into two; and the fact is, we don't yet know the spatial controls that decide when this is going to happen. Perhaps there is a clue to this and many problems of development somewhere through the looking-glass: how can my body have grown both a left hand and a right hand when all the aminoacid molecules in its proteins are left-handed? Geometrical self-assembly should fit together parts of either left or right handedness to give a whole of one handedness, and the term self-assembly is conventionally used, both in developmental biology and increasingly in chemistry to mean that kind of geometrical fitting. When the macroscopic patterns of development defy the inevitable consequences of self-assembly, the mechanisms of their formation must belong to some broader class of ways in which symmetry-breaking can arise. It is in that wider sense that I use the term 'self-organization'. The most obvious way for a system to escape from the dictates of geometry is to use mechanisms dependent directly on dynamics, such as the rates of reaction provided by the catalytic activities of enzymes, and thus only quite indirectly on the structural features that give those enzymes their activity. That this type of mechanism can account for the loss of one handedness on the molecular scale has been known since 1932 (see the introduction to Part II). By the same token, it can take particles of a single handedness and generate both right and left hands. This book is about how that kind of self-organization may work, and what evidence there is that it does in diverse instances of biological development.

My interest in working on development was triggered by the doctoral research of Thurston Lacalli on the alga *Micrasterias* (see Chapter 5). In his thesis (1973a), he gave an analogy (following an idea of J. Needham) for the two contrasting ways to go about investigating development, which I further elaborated in my 1993 book. It concerned the study of a Swiss watch to discover how it functions:

> One may take the watch apart and examine, list and diagram the springs, gears, shafts and so forth, and how they fit together. Yet one does not have a full explanation without the application of equations of motion to the whole. These involve concepts of momentum, moments of inertia, and simple harmonic motion arising from a restoring force proportional to displacement.
>
> If in the light of such a study . . . one were to set up a team to examine some other oscillating system of unknown contents, one might designate some people to take it apart and describe its contents in ever greater detail, and others to tackle other questions: what is the displacement that produces a restoring force, and what is the origin of that force? To be sure, these two parts of the team should exchange information, and the whole team is needed to produce the whole story.

In this analogy, the Swiss watch does not stand for a fully formed living organism. All its parts symbolize the genome and its products that are in active use during development to make patterns. And the oscillations of the watch movement correspond to the *spatial* periodicities of those patterns. But to appreciate what is in 'the whole story' one needs to consider what one is going to do with that story. What parts of it must one give most attention to if one wants to know: how to make a Swiss watch; how it works; or how to design some quite different oscillatory system. For pattern-forming systems, this book is about how they work, and how one tries to find out how they work.

1

Organizer, organize thyself

This book is about the great mystery of how living organisms develop the shapes and proper relative positions both of the whole organism and of all the parts that make it up. These are large-scale structures, ranging from parts of a cell visible by optical microscopy up to gross anatomy. My primary interest is not in the structures themselves, most of which remain after life has departed and are indeed commonly described from studies of dead material. Nor is my attention given mainly to structural details on the molecular scale within the provinces of the biochemist and molecular geneticist, details that could be regarded as the ultimate limits of anatomy, cutting-up that has reached the atomic scale. I am concerned mainly with how the tiny, partly fragmentary and partly one-dimensionally ordered genetic beginnings are transformed into the three-dimensional organism by the processes of development. This book is about processes, as they occur during life and make up a large part of what distinguishes living from inanimate matter.

The proper study and description of processes may involve diverse branches of the physical and chemical sciences (Harrison 1993, Chap. 8), but must quite often be primarily based on dynamics. That is the kind of explanation of developmental events to which I have devoted most of my effort, and is the main topic of this book. The postulation of ways in which diverse and complex shapes can readily be generated by dynamics is as old as the double helix model of DNA (Turing 1952). But the study of developmental dynamics still attracts only small numbers of researchers. Therefore this book is more about unanswered questions than about established explanations. My only certainty, for this as part of a generality for most fields of science, is that processes must in the end receive as much attention as structures if we are to understand the workings of the universe and all its parts.

Visitors to the increasingly fuzzy three-way intersection of science, pseudo-science and science fiction often wonder what life is like if it exists somewhere else in the universe, in forms that have originated quite independently of those on Earth. Is the DNA–RNA-protein molecular basis of all known life a unique set of building materials that will inevitably be found wherever there is anything that can be recognized as alive? My guess is that the essence of life does not reside in such molecular detail, indeed not in anything structural, but in processes and their dynamics, most particularly those that make up the development of the organism, its maintenance by assimilation of foodstuffs and excretion of waste products, and its reproduction. When Isaac Newton looked at the skies, he envisaged and established laws of motion that are the same for little things on Earth and huge objects in space. He and his scientific successors had no philosophical scruples against applying these laws to objects of unknown composition, and establishing that the laws worked very well to explain an enormous diversity of motions. They remain the primary basis of modern science, and explain the motion of the moon equally well whether its mass is made up of green cheese, Dante's nuns who failed to keep their vows, or a glob of interstellar rocky garbage. The cosmologist expects Newtonian concepts of motion and gravity and their Einsteinian modifications to apply to everything, including dark matter of still unknown composition. (Some readers of a philosophical bent may perhaps have the word Newtonian attached to a concept called 'clockwork universe', regarded as inadequate and out-of-date. I am not here concerned with that. The application of Newton's laws to whatever tiny, little or big bit of matter one happens to be considering is as timely today as when he formulated the laws.)

Does biological development yet await its Newton despite half a century of non-linear dynamics, just as evolutionary ideas had to await Charles Darwin despite the thinking of such 'transformists' as Lamarck, St-Hillaire and Erasmus Darwin up to half a century earlier? I hope not. The history of the evolutionary concept involved a most unusual sequence. The first step beyond data in the scientific method is supposed to be the formulation of a type of law defined as generalization from data. That the evolutionary succession is a valid generalization was recognized by some, but not generally accepted until Charles Darwin provided a mechanism for how it can happen. That is the second kind of law, theoretical explanation, and is supposed to arise out of already accepted generalizations.

In biological development, there is now quite a different situation. There is a plenitude of both data and theory; but they seem to be looking at each other from opposite sides of an apparently unbridgeable gulf. This is not at all what was expected by D'Arcy Thompson (1917, 1942, 1961). His title, *On Growth and Form*, clearly implied a preconception that form was a manifestation of the processes of growth, and that this would soon (after 1917) become a major direction for the advance of biology. Part of his basis for this was that he thought the possible limits of microscopy to have been almost reached. (The electron is to blame for many aspects of our lives today.) In the event, twentieth-century biology became in great part the quest for nanoanatomy, if I may be excused for using such a word, in the forms both of electron microscopy and of the molecular structure of the genome. The spectacular advances in this field have quite overwhelmed what could have been (and I hope some day will become) other large fields of biology much closer to the spirit of D'Arcy Thompson.

The developmental biology I discuss in this book can be categorized in a sub-discipline often called 'pattern formation'. But I do not distinguish between pattern and form, nor in any clearly defined way between pattern formation, morphogenesis, epigenesis and indeed most of biological development. All are concerned with essentially the same thing: the unfolding in space of the organization of the organism, something that is not in the first instance spatial, nor intrinsically structural. The essence of living organization is that a multitude of processes start running in sequences and in relationships to each other that permit their mutual support and minimize mutual destructive interactions with each other. But as part of the strategy of both necessary separations and proper couplings of these processes, living organisms develop significant shapes that announce pictorially the existence of this organization of continuing events. Spatial compartmenting is one of the most powerful methods used by organisms to convert a complexity of processes into a set of mutually interacting but semi-independent simplicities. Each of these simplicities is a pattern; and their mutual organization on many levels is a multitude of patterns.

Here, I am using the word 'pattern' to refer primarily to an abstraction in our minds, a pattern of interaction of processes. We may convert this into a picture that Nature has not drawn in the organism – for instance, a diagram of biochemical cycles. But Nature has gone beyond this, using the biochemistry as an artist to paint a self-portrait, the shape of the organism and its parts, which we can often appreciate statically as a pattern in the pictorial sense.

Is this, the obvious appearance of the biological world that everyone knows, perhaps more to be thought of as the abstraction, where the hidden but essential and down-to-earth reality is actually the set of processes generating the shapes? Of course, living shapes are most easily seen in very practical terms as enabling the organisms to run, swim, eat, use sunlight, extract nutrients from the soil and so forth. But all of these functions are ultimately ways that the organism goes about maintaining the organization of its processes of life. Spatial, pictorial pattern is a showing forth of the existence of these processes. The German word corresponding to pattern, 'muster', is derived from the Latin 'monstrare', to show.

But the English word is derived from the Latin 'pater', father, and the meaning of 'pattern' up to the sixteenth century was rather similar to 'patron'. This raises the question: when organization is evident, where may we find the Organizer? Formulations of this question go back at least to Aristotle, who wrote in his *Generation of Animals*:

> It would appear unreasonable to suppose that anything external fashions all the individual parts, the viscera or any others, because unless it is in contact it cannot set up any movement, and unless it sets up a movement no effect can be produced by it. Therefore there is already present in the foetus itself that which is either an integral part of it or separate from it. To suppose it is some other thing, and separate from it, is not reasonable. If it were, the question arises: when the animal's generation is complete, does this something disappear, or does it remain within the animal? We cannot detect any such thing, which is in the plant or animal and yet is no part of the organism. (Translation slightly modified by Harrison and Holloway from that of Peck 1942)

This quotation is from a few pages of Aristotle's work that are commonly referred to as indicating a belief in epigenesis (*ab initio* generation of its shape by each new organism) in contrast to preformation (pre-existence of the shape in miniature, e.g. the idea of a 'homunculus' within the human sperm). Aristotle did not use the word epigenesis, which was a nineteenth-century invention of von Baer. But the question in the quotation is very clear, and is the earliest statement I am acquainted with on the problem of self-organization. How can the organizer and the organized be identical? Twenty-three centuries after Aristotle, this is still being asked. For instance, in a review of wing and leg morphogenesis in the fruit fly *Drosophila*,

Serrano and O'Farrell (1997) restate essentially the same question in the words: 'One of the most mysterious features of morphogenesis is that structures form themselves as they grow … How are the actions of morphogens coordinated with the process of growth?'

Both the ancient and the modern formulation of the question contain, however, what I believe to be the germ of the answer. Aristotle's references to 'setting up a movement', and Serrano and O'Farrell's use of the words 'form', 'grow', 'actions' and 'process' direct our attention towards dynamics as the field of work that should lead to the desired explanations. My purposes in writing this book are two-fold: to indicate some of the diversity of developmental phenomena; and to describe those advances in dynamic theories of pattern formation, mainly from 1952 onwards, which I believe to be the first few steps along the road towards discovering unifying principles underlying the experimental diversity.

This book may be read for general interest. But I hope also to encourage more scientists of all backgrounds to take active part in the quest for these unifying principles. To those with physical and chemical backgrounds who are accustomed to studies on clean samples containing no more than a handful of different substances participating in one or two reactions (or none at all), I suggest that they should not be afraid of systems containing hundreds of thousands of chemicals. If such a system is alive and in good health, it is not dirty but indeed almost the ultimate in cleanliness. The reacting systems are so organized that they not only avoid interfering with each other but actively assist each other and defend each other against contamination. To biologists who are accustomed to meticulous description of detail in diverse organisms, I suggest that they should not be so completely preoccupied with getting the details right as to neglect the presence of new unities to be found. Much of this search involves highly selective use of items from the available genetic and biochemical knowledge pool, rather than any attempt to put all of it together into a single computer program.

1.2 BREAKING OUT

For anyone setting out upon this quest, I present the painting on the cover as an appropriate heraldic blazon. The original hangs in a hall of the University of British Columbia library. It is *Emergent Image* by the late Jack Shadbolt, and was described by the artist as representing a butterfly breaking out of the cocoon. The picture therefore has the primary characteristic that must be shown in any good description of development: it moves. Many pictures of biological pattern, whether made photographically or by an artist, are derived from the study of dead material fixed for microscopy, and so convey to the viewer the cessation of movement that is death or fixation. But this picture is of pattern formation. To me, it shows much more than the emergence of a preformed structure from an external restraint. It strongly evokes the emergence of an adult organism from the whole sequence of development within the earlier stages of the organism.

Clearly, the motion in the picture is from left to right and involves increasing size. But the starting motif on the left, a set of wiggly lines, is not a molecular microstructure such as the DNA in an egg. Rather, it remarkably suggests a type of elementary spatial pattern that has attracted attention in biology, chemistry and theoretical biophysics, all at dates later than Shadbolt's painting. That is, a set of parallel stripes, one of the simple motifs for patterning in 2-D. Shadbolt shows wavy stripes. That waviness indeed turns up in computations from stripe-forming dynamic models (Lyons and Harrison 1992a; cover picture of Harrison 1993), in non-living chemical pattern-forming systems such as the CIMA (chlorite–iodide–malonate) reaction (Ouyang and Swinney 1991), and in at least one living pattern that shows clear indications of dynamic generation (Kondo and Asai 1995; this book, Fig. 9.8).

The motif is repeated in the striking pattern of white and dark stripes in an elongated oval. This strongly suggests the first action in an insect egg that gives rise to the body segmentation characteristic of the Insecta (Chapter 8). Nature does its painting of the earliest embryological events with particular RNAs and proteins. But it is a recurrent theme of this book that Nature's palette is huge, and that the same kinds of painting processes can be done with very diverse materials. Striped patterns can appear from the earliest signs of segmentation in an embryo, to postnatal organization of the visual system in a mammalian brain (Fig. 9.7), to the coat pattern of a zebra. Examples are multitudinous; but are the modes of formation of any one category of patterns equally numerous, or are there just a handful of them, or is there just one, as Newton's laws of motion apply to everything? My inclination is to the view that the answer will finally be 'a handful', but that to seek them it is going to be most useful to have the unifying idea in mind.

And then Shadbolt's picture explodes to the right. This clearly indicates breaking out of the cocoon, but I would add 'amongst other things'. There is an ambiguity in the middle space. The fragments from the left seem not necessarily to be falling into discard, but just as probably to be moving to the right, to build the butterfly. Strength of movement is well shown, but very little of detail in organizing the organism. This book is about how to start trying to fill in that great gap in our knowledge.

The artist can paint pictures that tell a story of movement. But much of the information to which a biologist must devote extensive attention is from dead and fixed material and quite lacks that artistry. In my previous book (Harrison 1993, Chap. 1) I mentioned the story of the conductor Franz Lachner, who complained of the score of Wagner's *The Flying Dutchman* about 'the wind that blows out at you wherever you open the score'.

I pointed out that only an expert musician can feel this wind when looking silently at the printed score, and suggested that biologists need to acquire a similar expertise in seeing motion in electron micrographs. Subsequently, I have become involved in a project on the possible mechanism of some rather spectacular pattern formation that happens during chromatin condensation in sperm of some species. This is described in Section 7.2.2 of this book. It has given me direct experience in trying to ferret out time-sequence information from electron micrographs, in which each one was necessarily made from a different killed specimen. The experience has only reinforced my previously expressed view.

But it has become commonplace to have continually moving pictures in any ordinary room. The natural habitat of these organisms is the computer screen, and they are classified as screen-savers. I can't stand them. When I am not using my computer, it sits there with a screen as uniformly dark grey as volcanic ash before life has moved back on to it. This is because I find screen-savers immensely distracting. I want to watch them and work out what is going on. At one time, I had a screen-saver of tropical fish (of diverse hitherto unknown species, I think) wandering across the screen, sometimes reversing direction, disappearing at all four boundaries and sometimes reappearing at the same point or on the opposite side. This show was rapidly destroying my scientific endeavours; I had quite a lot of statistics compiled on the movements of these fish, and was approaching questions of whether each one had a continuous existence or whether they were created and annihilated, and so forth, when a hard drive crash finally rescued me. The point of this story is that I had no interest in the details of the program that was running this show. All I wanted to know was what things were in the algorithm that the program was written to implement; and I had hope of working this out from observations.

My attitude to developmental phenomena is much the same. If an organism has anything close to an equivalent of a computer program, it is the genome, a mass of detailed statements in which it is quite usual for a single trivial error to have catastrophic consequences. The programs that people in my research group (especially Stephan Wehner and David Holloway) have written for computations of simultaneous three-dimensional chemical patterning and growth for branching processes in plants comprise about 20 000 lines in the C programming language. A single error in one line can have consequences as bad as those of a mutation in a genome.

What is the way to go about studying how the developmental event happens: to go for the genome by studying mutations, which is like trying to write the computer program from the effects of diverse errors made without the observer knowing just what they are? (Great advances in genetics have been made this way, before there was any way of making known chemical changes in the genome.) Or, to study larger-scale features of the event and its unfolding in time, and to try therefrom to write the algorithm, a much briefer thing than the program? My answer to this question is that both must be done. It is best to start building a bridge from both ends.

1.3 KRESPEL

The early nineteenth-century stories of E. T. A. Hoffmann are best known from the use made of three of them in Offenbach's opera. One of these concerns a Councillor Krespel and his daughter Antonia. In the original story (translation, Kent and Knight 1969), much more is told of the eccentricities of the father. To construct a new house, he had the masons build, on a square foundation, four walls with no apertures. He had them halt when they had reached a particular height, and then specified, one by one, the places for material to be knocked out for windows and doors, and the exact size of each of these. There was never any architect's plan. 'Everything had to be done on the spot, according to the orders of the moment. In a short time a completely finished house was standing, presenting a most unusual appearance from the outside – no two windows being alike, and so on – but whose internal arrangements aroused a very special feeling of ease.'

This episode can surely be read as a good metaphor for one of the ways that living things go about constructing themselves, in contrast to the conventional architectural method of following a plan by leaving all the necessary gaps in the structure as it is built from the ground up. An organism is completely bounded from the start of development, and the processes continually take note of these boundaries, as Krespel kept on looking at the walls to see what he would do next. Because of this kind of interaction with what is being built, Krespel, though he is looking at the house from outside, is metaphorically close to that Organizer that is part of what is being organized and yet cannot be part of it, the concept that so puzzled Aristotle.

By writing, above, 'one of the ways', I imply that there is a restriction on the generality of the metaphor. This restriction is essentially to developing organisms that do not yet have parts irrevocably committed to specific fates, but are still completely capable of 'regulation' (in the biologist's sense of that word). In the animal kingdom, amniote embryos have this capacity when they have, e.g. as many as 20 000–60 000 cells in the chick embryo (Streit *et al.* 2000). In the plant kingdom, development continues throughout life. The building of a flower, comprising many events of pattern formation, morphogenesis and differentiation of cells and organs, starts only at the culmination of the life cycle, when a meristematic region becomes specified as floral, and Krespel arrives to take charge.

But a major inadequacy of Hoffmann's metaphor is that the boundaries of the house do not expand. All of Krespel's eccentricity could not enable him to start with the walls of a bungalow and end up with a cathedral. Likewise, the greater part of the work on dynamics of biological pattern formation to date has been done for regions of fixed size, shape and boundary conditions. This restriction is stated explicitly, for instance, in Meinhardt (1982) and Nicolis and Prigogine (1977). And it is evident from the contents of those books that an enormous amount of work highly relevant to how biological development happens can be done without stepping outside that restriction. Indeed, many pattern-forming events take place in embryos that cannot feed until those events have happened, and cannot grow until they can feed.

There are other instances in which one cannot avoid the fact that the four walls of Krespel's house should be continuously growing as he chooses places for windows and doors. These include nearly all the events of plant development, and much of later animal development, such as the formation of limbs, multilobed glands, lungs, etc. For these, the question of coordination of pattern formation and growth must be addressed. In what follows, I try to emulate Krespel's level of eccentricity in two ways. First, I deal with this more complex kind of development before treating patterning in regions of fixed size and shape. Here, the method in my madness is that I want to raise questions as much as to provide answers, and phenomena of simultaneous patterning and growth are best to draw out a good map of question-marks.

Second, I concentrate first on plant development, recommending it as a highroad into most of the territory on that map. Though the biology of our own kingdom is inevitably a main preoccupation of the human species, many generalities of biology have been discovered first in plants. Therein, the cell was first seen by Robert Hooke, long before it was recognized as a component of all life. The nucleus also was seen first in plants; and where would our knowledge of genetics have reached today without the old plant-breeder, Gregor Mendel?

But back to Krespel. When he asked for things to be done in what seemed to be a strange order, he was nevertheless meticulous for each new task in specifying precise dimensions, stopping wall-building when it had reached the height of a two-storey house, and giving exact sizes for each door and window. This is a common feature of biological development. The components of living action, chiefly chemical reactions, transport processes and physical forces, are capable of measuring distances without the intervention of human intelligences and their elaboration of measuring methods. When I get to my favourite topic of chemical-kinetic mechanisms, I show that this property is capable of quite straightforward explanation. (This was treated with somewhat more mathematics in my earlier book, Harrison 1993). Again, the main point is: does any postulated mechanism possess this property, when the biological events being considered clearly have it?

In relation, however, to the general strategy of explaining how development controls itself, the significant feature of Krespel as a metaphor for the self-organizer is that he made his judgements of size and position in relation to a larger structure already grown. That aspect of development is a major theme of this book. A pattern is an entity occupying all or part of the space of an existing entity. Where, then, do we find the metaphor for the much more popular approach of tracing developmental events from the spatially tiny store of information in the genome? That, surely, is within Krespel himself. His eccentricity of method and piecemeal decision-making deny the existence of a preformed architectural plan. Likewise, the genome holds the potential for production of substances without a preformed plan of where they are to go. But a seemingly planned large-scale structure takes form. The dynamics of the planning are within the organism, and their reality no less than that of the structures to which they lead.

Aristotle was not aware of the existence of DNA; yet he said quite clearly that DNA is not the Organizer that we are to search for: 'We cannot detect any such thing, which is in the plant or animal and yet is no part of it.' DNA is surely a part of the organism and indeed, as is well known from its forensic uses in precise identification of individuals, it can be a part of long-dead detached fragments of the organism.

The late Paul Green, who instigated the writing of my previous book (Harrison 1993), wrote to me in the late 1980s: 'The idea of coupling one developmental change to one section of the genome is inadequate because there are far too many developmental events.' At that time, very few people seemed to be expressing such views. But the discovery in the Human Genome Project that the library of information we carry in each of our cells is only about one-third as large as most people had supposed, and no more than three to six times the size of the library that a bacterium carries, is making such views much more popular. We should remember that Paul Green said it first, long before the Human Genome Project started. He continued:

> The solution to this information paradox is that an organism inherits rules that spell out the progression … The rules are, or are like, time-based differential equations which have the ability to encode complex sequences with high efficiency. Thus one has to regard development as an integration through space and time, the genome providing the equivalent of the differential equations. Thus there is no escaping the calculus when studying development.

His last paper (Green 1999) was entitled 'Expression of pattern in plants: combining molecular and calculus-based biophysical paradigms'. His use of the term 'calculus-based' may seem a little quaint; it has done to some of my acquaintances. But even so dedicated a promulgator of meticulous detail in describing development as Scott Gilbert, in his textbooks, has said essentially the same thing: 'However, this does mean that some of us will have to learn calculus' (Gilbert and Sarkar 2000). Gilbert's writing is of descriptions of events, and Green's of mechanisms, in the natural sciences. Calculus is a branch of mathematics, not natural science. What is the meaning of their usages of the term, and also of Green's 'to encode sequences with high efficiency'?

A physicist, Michael Wortis of Simon Fraser University, has asked a question that perhaps serves as a succinct answer to this one (Wortis *et al.* 1993): 'We all know that biological systems do marvellous things.

Are they marvellously complex machines or, rather, simple machines with marvellously complex behaviour?' Surely the answer should be 'both', and the latter is what Green was postulating. But the former has attracted overwhelmingly the greater amount of attention in the past half-century, and the latter has yet to come into its own (and the 'complex behaviour' is the aspect that needs calculus to take it from obscure mystery to intelligibility).

This is not the sequence of events that D'Arcy Thompson (1917) anticipated when he wrote that 'the things that we see in the cell are less important than the actions which we recognize in the cell'. He could hardly have foreseen that the advent of the electron microscope and the advance of molecular structure determination would together lay down a highway along which the momentum of descriptive biology would roll to the ultimate limit of anatomy, the atomic structure of the genome.

What is the other way, so far an intermittently flagged tentative trail through the forest of information? My compendium of metaphors is leading me to two places, Newtonian physics and chemical kinetics. Both of them in particular senses ignore trees to penetrate forests. Newton's physics established the unity of interactions leading to motion between all objects quite regardless of their detailed structure. Chemical kinetics, my home discipline, commonly deals with mechanisms having dozens of steps, in which a spectacular kinetic property requires only two or three of the steps to explain it. This might be, for instance, the observable property that the reaction to form water in a mixture of hydrogen and oxygen gases occurs explosively. The simplest approximation to a more or less complete mechanism for the reaction (Kondratiev 1969) has 13 steps involving the species H_2, O_2, OH, H, O and HO_2, with a footnote that there are probably a few more steps involving H_2O_2. But it is a reaction of branching-chain character, in which some steps generate two reactive radicals out of one radical and a molecule of ordinary reactivity. These steps are only two in number:

$$H + O_2 \rightarrow OH + O \text{ and } O + H_2 \rightarrow OH + H.$$

These two by themselves, and without any of the others, explain the explosive nature of the reaction dynamics.

Once upon a time, in an undergraduate course on physical chemistry for chemical engineers, I had taught this sort of standard textbook account of explosions in hydrogen–oxygen gas mixtures. Then I thought I would go a step beyond the textbook and do something similar for the other best-known gas explosion, methane–oxygen. Among many known steps, I had quite a hard hunt to locate the two or three that gave the mechanism that 'branching-chain' property. But I found them in the end because I knew what I was looking for. In trying to cope with many-step mechanisms, it helps to know what sorts of 'leitmotifs' in two or three steps will give the mechanism particular properties. That is my attitude to Turing reaction–diffusion models: that they constitute a leitmotif to be borne in mind when one is struggling with such things as enormous networks of gene and protein interactions.

As a phenomenon, an explosion is perhaps not very close to biological development; but oscillatory reaction clearly has some relevance to spatial pattern formation. One of the best-known reactions of this type is the Belousov–Zhabotinsky reaction, oxidation of malonate by bromate with a cerium (III) catalyst. The mechanism of Field *et al.* (1972) for this reaction has about 30 steps, and even that number does not complete the oxidation from malonate to carbon dioxide. But to explain why the reaction proceeds in an oscillatory fashion, one needs to pay attention to about three of these steps.

By the same token, to explain how an assembly of biochemical processes generates spatial pattern, it should not, I believe, be necessary to attach particular significance to every process in this multitude. Although one approach that is gaining popularity, now that so much genetic information is available, is to program a computer to simulate the dynamics of a complete genetic network, this is not the only approach that is likely to be useful. Hunting through a network to find a small number of reactions that together have pattern-forming dynamics could be equally useful. And, just as Newton's laws of motion apply to things large and small, so the dynamic motifs of a small number of interactions that can generate pattern are not necessarily restricted to chemical reactions between molecules. Sometimes, the interacting objects might be biological cells. (See the topic 'cell as molecule', touched on in half a dozen otherwise diverse topics in Harrison 1993.)

1.5 THE ORGANIZER OF WHAT?

I opened this chapter with a rather broad-ranging reflection on the meanings of the word 'pattern'. But the book addresses a much more precisely defined topic: the search for mechanisms by which living organisms control the production of their various parts in the proper shapes and sizes; and particularly, the instances in which this control is likely to be dynamic, rather than a structural fitting together of parts in building-block or jigsaw-puzzle manner. (The latter is what I understand to be meant by the term 'self-assembly'; 'self-organization' includes all possible mechanisms.)

Except for this restriction of looking for instances of dynamic control, when I mention 'various parts' of an organism I mean any and all parts, and most particularly those concerned in the crucial early stages of embryogenesis and organogenesis. This is not at all to belittle work on more obvious and more easily studied patterns because they are superficial. Important work in relation to dynamic mechanisms has been done on animal coat patterns (Murray 1981a, 1981b, 1988, 1989), sea-shell pigmentation patterns (Meinhardt 1995) and surface patterns of tropical fish (Kondo and Asai 1995).

To try to clarify where my topic fits into the conceptual framework of developmental biology, I show as Fig. 1.1A what appeared as Fig. 1.1 of my previous book (Harrison 1993), juxtaposed with, as Fig. 1.1B, a much more recent schematic by Bornholdt (2005). Both of us are recommending hunts for simple extracts from the complexity of living systems on which one can do important mathematical or computational work. The strategy of my approach is defined by the arrows at the top, starting from the macroscopic study of morphogenesis and seeking primarily the type of control rather than the molecular nature of the substances involved. I distinguish, from a physicochemical viewpoint, three categories of mechanism: kinetics, approach to thermodynamic equilibrium and structural self-assembly, and expect mechanisms within all of these to play some part in self-organization. Bornholdt's approach starts from the plethora of known facts at the molecular level that is now getting to be available for large genetic networks, and categorizes ways of simplifying them to something tractable in computations. This is likely to become the most popular way of moving from genomes and proteomes to functions, including the events of development. But I believe that, especially when the kinetic category of mechanism seems the most promising, my approach may be a necessary way to go hunting for developmental leitmotifs in the large networks.

Types of pattern for which mechanisms in this category, especially reaction–diffusion, have been explored include localized single organs as well as patterns of repeated parts. For the latter, the pattern repeat distance can usually be identified with a wavelength in the mathematical treatment. A striking feature of such repeat distances is that, at the time of first morphological appearance, they most commonly have similar values – of the order of tens of micrometres

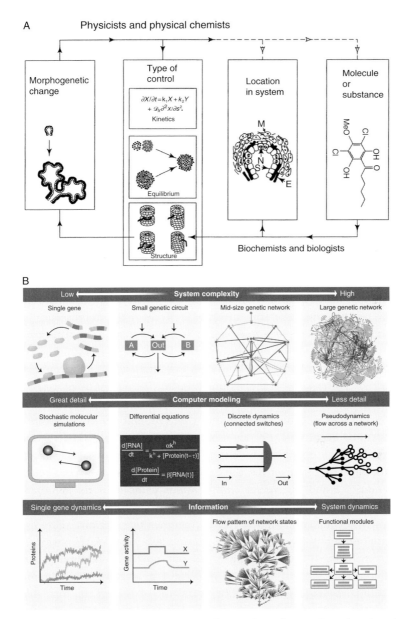

Figure 1.1 Contrasts in strategies of research. A: Fig. 1.1 of Harrison (1993), the arrows at the top define the approach to mechanisms of development from macroscopic studies, while the arrows at the bottom show the approach starting from known molecules. Boxes 1 and 3 (from left), show development of a lobed gland, from Bernfield *et al.* (1984), see also Fig. 9.13. In box 2, 'equilibrium', cell sorting, see also Fig. 7.2; 'structure', the self-assembly of the tobacco mosaic virus. B: from Bornholdt (2005) with permission, a schematic of various degrees of simplification of a large genetic network to find within it mechanisms for events.

(µm) – for very diverse events in organisms that achieve a very wide range of adult sizes, e.g. hairs or lobes as parts of single-celled algae of complex shape (Chapters 3 and 5), stripes of gene expression as the beginning of segmentation of an insect (Chapter 8) and human somites as they form in a 21–8-day embryo (Section 9.3). For a reaction–diffusion mechanism, the wavelength is strongly correlated with the rate constants of the reactions, and hence with the time taken by the pattern-forming event. Similar patterns, but with very different quantitative time–space relationships, may arise by very different mechanisms. An instance is my suggested mechanism for formation of lamellar patterns of chromatin in sperm nuclei, with spacings of only tens of nanometres (nm), 1000 times smaller than the usual spacings in biological development, but apparently forming quite slowly. I ruled out reaction–diffusion but suggested a mechanism of phase transition known as 'spinodal decomposition', which does not involve chemical reactions and completely lacks the time–space correlation of reaction–diffusion (Section 7.2.2; Harrison *et al.* 2005). The events I more usually study require only the lower magnifications of the light microscope, but always with an eyepiece micrometer in place; spatial quantitation is crucial to discussion of mechanism.

Not all mechanisms of one kind are equally good at forming all kinds of pattern. Within reaction–diffusion, generally categorized as Turing mechanisms, the Brusselator of Prigogine (1967; Prigogine and Lefever 1968) is good for repeating patterns, but the mechanism of Gierer and Meinhardt (1972) is much better for localized single organs. This contrast is illustrated here by the accounts of computations using a Brusselator mechanism for the growth of a star-shaped algal cell in Sections 5.2 and 5.3, versus computations using a Gierer–Meinhardt mechanism for the positioning of a vertebrate heart in Section 9.1.

When I first designed Fig. 1.1A more than a dozen years ago, I envisaged physical scientists as most likely to take the approach starting from the macroscopic scale. But now it is becoming quite common to represent individual molecules in numbers like hundreds of water molecules and smaller numbers of solutes in molecular dynamics computations of various phenomena. Reaction–diffusion computations commonly use concentrations of substances as the variables, and look like the chemical kinetics of the 1930s and 1940s. A useful perspective is to consider a 10 µm cube as a unit of which a number need to be put together to make the usual size of a pattern-forming domain. If this cube contains an aqueous solution of a highly reactive substance at 10^{-9} M concentration, then it contains about

3×10^{13} water molecules and 600 solute molecules. The latter, multiplied by the number of cubes one needs to assemble for the pattern-forming domain, is a good estimate of the smallest number of molecules one needs to cope with for any substance in a computation. This has two important implications: first, a pattern-forming event is beyond the current scope of molecular dynamics computations; second, numbers of the most active molecules in a mechanism can be small enough that Poisson errors in concentrations could be a few per cent. It is then necessary to consider how developmental mechanisms go about suppressing the accumulation and propagation of such errors in sequences of developmental events. This topic is discussed in Section 8.3.

1.6 DESCRIBED WHERE? (THE ORGANIZATION OF THIS ACCOUNT)

In my previous book (Harrison 1993), I wrote that 'all the words lead towards the equations, and the words are really useful only to people who are going to follow them that far'. But I then sought to contradict myself by entitling a chapter 'Pictorial reasoning in kinetic theory of pattern and form'. In using the term 'pictorial reasoning' I had in mind that many chemists do a lot with sketches of the 'shapes' of such things as s, p and d orbitals, and σ and π bonds, when they are not at all the kind of chemists who intend to spend a lot of their time with Schrödinger equations.

The present book is not intended to lead inevitably towards equations. I have tried to approach the whodunit, the identity of the Organizer, serially, with a new clue at the end of each episode. For those seeking a compendium of the clues:

(1) Whoever the Organizer or Organizers may be, they did something somewhere, and the first thing the detectives do is, commonly, to measure things up at the scene of the crime. I take up this topic in Section 2.3, 'Numbers of parts, spacings, wavelengths', thereby maintaining that one of the most significant properties of a pattern-forming mechanism, which any good theory of such mechanisms must explain, is the ability of the Organizers to set up quantitative scales of distance.

(2) The detectives are then likely to go looking for circumstantial evidence with a magnifier, a phrase I have therefore used in

the heading of Section 3.1.2. There, I show that behaviour looking very much like the physical chemistry of enzyme kinetics, but for yet unidentified substances, can be found in measured spacings in a pattern.

(3) The character of the Organizers is further revealed by the patterns they make. They seem to be artists who treat space as a musician treats time, by generating periodicities. These may be waveforms in one, two or three dimensions of space. I do not deal with them in that order. The 2-D case of patterns as they would appear at any instant in the vibrations of a circular drumskin is mentioned first, in Section 3.2. There is a good reason for this. Many of the events of biological development, whether for animals or plants or single cells, occur in thin layers, e.g.: single layers of cells (epithelial sheets) in the insects, and the tunica layer in plants; pairs of such layers in vertebrates; and cell surfaces, perhaps together with the cortical layer of the cytoskeleton, in single cells. Both 1-D and 3-D waveforms are taken up in Chapter 4. For waveforms in all dimensionalities, I draw upon well-known mathematics of their shapes; but what is needed can be understood pictorially, without looking at any equations. But how did the Organizers make them? As at about this point in any conventional whodunit, I show the reader that we don't yet know much about the culprits we seek. Are their methods those of mechanical distortion or chemical action? In Section 4.4 I start to approach the latter possibility, my major topic, by describing some features of the growth of a chemical concentration wave, again just pictorially.

(4) The problem now arises that these nefarious artists have developed a revolutionary technique in which they continually bend and stretch the canvas while they are painting a picture. Chapter 5 shows our sleuths going for this part of their method with computers, but still defers any account of how they do the painting.

(5) The methods of the gang known as the Turing morphogens are at last revealed in Chapter 6. Amongst all other violence in the story, a historically based whodunit has to do violence to history. In this one, it is shown that a guy called Wigglesworth (regarded by many biologists as a friend, while they greatly suspect Turing) actually started the whole thing in 1940 with the 'inhibitory field' concept. Once one lets the activator also

spread, albeit less than the inhibitor, and the patterning power of the mechanism is enormously increased. This is the snare that instantly enmeshes the readers in second-order non-linear partial differential equations.

The above imaginary component of a novelette is written, however, only to show a trail leading to the Turing morphogens. While I am convinced that the Turing dynamics will eventually come to be recognized as of major importance in biological development, I do not ascribe to them a unique and universal significance like that of Newton's laws of motion, or of DNA in genetics. In my previous book (Harrison 1993, Chap. 8), I constructed a classification of developmental mechanisms, mimicking biological classification and using the major divisions of physical chemistry (kinetics, equilibrium, structure) as its phyla. In the kinetic phylum, I placed 'reaction–diffusion' and 'mechano-chemical' as two classes. I return to this in Chapter 7 of this book. In addition to mechanochemical theory, I discuss Paul Green's concept that morphogenetic events at a plant apex are generated by mechanical buckling, in which mechanics alone can potentially be pattern forming, without assistance from chemistry other than, of course, to provide the deformable materials. This is not a theory in the kinetic category, in which there is a continual flow of material through the system, keeping it out of thermodynamic equilibrium. Mechanical buckling can achieve patterning by approach to equilibrium. It is mentioned in connection with spacings in Section 2.3. An elementary account of the theory is in Section 4.2; and I give a perspective on the relative promise of reaction–diffusion, mechanical and mechanochemical concepts in Chapter 7.

Part I   Watching plants grow

Plant cells have a rigid cell wall. Therefore, unlike animal cells, they do not 'sort out', i.e. change their relative positions during development. Plant development is thus discussed largely in terms of two processes: cell division, and most particularly the control of its directionality; and change of cell shape by anisotropic extension of the cell surface, often in the form of elongation along the axial direction of a stem or root. In molecular terms, arrays of microtubules are of great importance in plant development, because they can define directionalities.

My account in the four chapters of Part I is not primarily concerned with those features of plant development, but with another very general feature of plant development, branching processes. These commonly occur in apical regions having circular symmetry, and demand explanation by mechanisms capable of breaking that symmetry. This is a significant guide to possible classes of mechanism to explore. But whereas such a consideration leads one towards generalization, relevant experimental data are usually for specific examples. At the present stage of partial knowledge, there is no unique way to put together an account interweaving experiment and theory into a fabric of scientific understanding. My attempt at a start on this has the following structure.

Chapter 2 is preliminary, introducing some interesting developmental phenomena. The only theory in it is the concept that a branching pattern is a waveform, an idea to which I return time and again in this book. Chapter 3 deals with whorls, usually of more than two organs, all formed simultaneously. The two specific examples treated in detail are the only ones on which I and my group have done any extensive biological experimentation. For *Acetabularia*, I present evidence pointing towards a reaction–diffusion mechanism. For larch embryos, I give evidence that the entire set of cotyledons arises as a waveform, for the origin of which physics, chemistry or both must provide some explanation. Chapter 4 is devoted to the simplest, very common and probably most ancient form of division of an apex, dichotomous branching. This chapter is essentially all about theory, introducing particularly 3-D waveforms on a hemispherical shell. But in Section 4.2 I try to give a perspective on the rival claims of mechanics and chemical dynamics to take centre stage in the story of branching. This sits comfortably in Chapter 4 because of the late Paul Green's potato chip analogy (Section 4.2), but in fact is relevant to much more than dichotomous branching. Chapter 5 has as its major theme an instance of repeated dichotomous branchings leading to parts all having coplanar axes, quite unlike whorled structures, and thereby introduces two big challenges to theorists using computer modelling: how to do it for a system that is continuously getting bigger; and how to do it in three spatial dimensions.

Phyllotaxis, the sequential formation of leaves in a spiral around the growing stem of higher plants, is a topic which quite fascinates many people, because of a curious relationship of the angle between successive outgrowths to the 'golden section' and hence to the Fibonacci sequence of numbers. From the viewpoint of searching for mechanisms of symmetry-breaking and hence of branching, however, I believe that the mechanisms for simultaneous and sequential processes will eventually be found to be, in essence, the same, but that it may be easier to find them in the first instance by looking at the simultaneous events. The main reason for this is that the mechanisms for branching should be essentially the mechanisms by which living things can set up quantitative scales of linear measurement. A ruler and the scale of distance it depicts are patterns of regularly repeated parts. The spatially quantitative experiments needed to show the existence of this ability and to make further tests on its properties are much more easily made on patterns of simultaneously formed parts that directly exemplify the ruler with equally marked out divisions, than on patterns produced sequentially, one part at a time, often with quantitative variation between one event and the next. The rest of Part I essentially elaborates the philosophy of this paragraph.

2

Branching
How do plants get it started?

2.1 BRIDGES, BRANCHES, PLANTS AND PROTISTS

At a developmental biology meeting, in front of a poster advertising somewhat generally the activities of my group on kinetic theories of pattern formation, I was asked by a plant geneticist: how do plant meristems manage to break circular symmetry? I was rather delighted to hear such physicists' terminology as 'symmetry-breaking' used in such a connection by a biologist. To my disappointment, I was unable to expand this incident into a collaboration; but for a moment this interaction was certainly right in the middle of the kind of bridge I keep trying to build. What such symmetry-breaking leads to is branching, the most common and well-known feature of plant growth, the thing that makes a bush bushy or a flower floral. And, despite the vast amount of information that is piling up on genetics, with all its powerful applications, we don't know what makes branching start, i.e. what makes apparently identical portions of a simple shape behave differently to make the shape more complex. Turing (1952) devised his theory of morphogenesis having in mind examples in the animal kingdom, especially the loss of spherical symmetry in a blastula when gastrulation commences. For plants, I return to the discussion of symmetry in my account of dichotomous branching in Chapter 4, with the idealized starting point of a hemispherical meristem topping a cylinder.

Part I is taken up quite largely with examples from the work of my group and describes also the approach of the late Paul Green. He and I shared a lot in our philosophical approach to biological development, and it was Paul's initiation and encouragement that led to the existence of my previous book (Harrison 1993). Nevertheless, our approaches to the problem of branching at a meristem or growing tip may seem very different. Paul liked to distinguish them as physics and chemistry; but

having always resided in a boundary region between those two disciplines, I was never really happy with that way of putting it. More precisely, Paul pursued theories in which mechanical forces have to be
taken into account explicitly, with pattern emerging from mechanical
buckling, while I have explored the capabilities of chemical kinetics to
generate patterned concentration distributions. These may seem like
very different approaches; but they have an underlying unity.

This lies in the preconception (Crombie 1959) or paradigm (Kuhn
1962) – I have always preferred Crombie's word, though Kuhn's has
become rather popular – that a pattern is an entity, and must be
considered as such in seeking the mechanisms of its formation. It may
readily be asserted against this view that what appears at first glance
to be a set of whorled structures, such as the petals of a flower, is
actually formed sequentially rather than simultaneously. But in pursuing reaction–diffusion theory, I have generally found that it takes only
minor modifications of a mechanism to turn it from a simultaneous
to a sequential generator of a number of repeated parts in a pattern.
The ability to do the former is an essential feature of the nature of the
mechanism, and I have certainly felt the urge to study experimentally
phenomena in which all the structures in a whorl appear at once.

These I have found in developmental events both at the growing
tips of very large single cells and in multicellular organisms. This raises
the question of what I mean by a plant, because it is popular to take all
single-celled species out of the plant kingdom and put them into the
Protista. Yet in the immediate subtidal zone of many warm seas, one
may find seaweeds that are each several centimetres in height and
furnished with whorls of hairs, so that they look remarkably like small
horsetail plants; but each is a single cell, with only one nucleus. These
are of the order Dasycladales (Greek, hairy branches), the best-known
genus being *Acetabularia* (Fig. 2.1A), named from a resemblance of the
cup-shaped reproductive cap to a Roman vinegar bowl, acetabulum.
(For the 2000-year history of work on the Mediterranean species, see
Bonotto and Berger 1997.) And in freshwater ponds around the world,
one may find organisms each big enough to be just visible to the naked
eye as a green blob, and revealed by low powers of magnification as
beautifully star-shaped structures, hence given the generic name
Micrasterias (little stars), and classified among the desmids (Fig. 2.2).
Again, each organism is a single cell. In what follows, I do not adopt
the 'protist' terminology, but describe both *Acetabularia* and *Micrasterias*
as plants, thinking of both as algae and chlorophytes. For the dasyclads,
this corresponds to the usage of Berger and Kaever (1992) in their

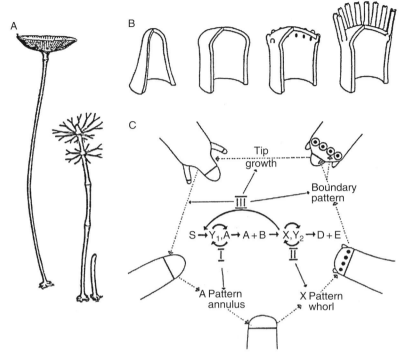

Figure 2.1 Whorl formation in unicellular *Acetabularia acetabulum*.
A: Three stages in development: the single nucleus remains in the rhizoid, at bottom; vegetative whorls form every few days and are deciduous (scars of two earlier whorls shown); the saucer-shaped reproductive whorl (left) appears once, at the culmination of the life cycle. Adapted with kind permission from Gibor (1966). B: Four stages in vegetative whorl formation, showing preceding tip flattening. With kind permission from Dumais and Harrison (2000). C: Feedback loops (Roman numerals) needed in a chemical mechanism to establish: I annular patterning giving tip flattening; II whorl pattern; III new boundaries confining pattern formation to new tips. With permission, from Harrison (1992).

monograph on the Dasycladales, which should surely be browsed by anyone seeking inspiration from Nature in their thoughts about patterns; as also should Couté and Tell's (1981) collection of scanning electron micrographs of the desmids.

In the foreword to Berger and Kaever (1992), Ralph A. Lewin wrote: 'The Dasycladales are among the most improbable unicellular organisms that exist.' They are indeed. Looking at an *Acetabularia* cell, at a fairly late stage in its life cycle, about 4 cm long and carrying several

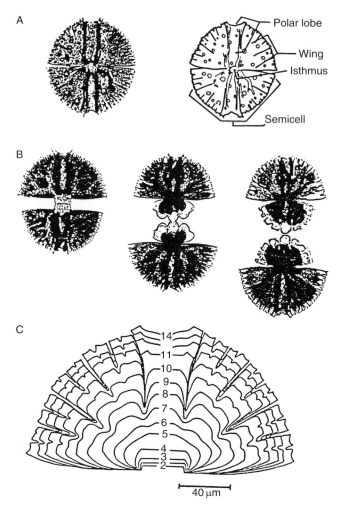

Figure 2.2 Semicell morphogenesis in the unicellular desmid alga *Micrasterias rotata*. A: Fully formed cell, sketch from photomicrograph and outline with terminology; B: Stages 3, 7 and 10 in development of two daughter cells, 60, 140 and 200 minutes after mitosis; C: Stacked semicell outlines at 20-minute intervals, showing the sequence of dichotomous branchings. All with permission from Lacalli (1973a).

whorls of hair (Fig. 2.1A, on the right), one might guess that it comprised about a million cells. Yet there is only one cell, with one diploid nucleus at the bottom end (in the rhizoid, a structure serving as holdfast to a rock in the immediate subtidal zone of a warm sea). At the culmination of its life cycle, *Acetabularia* makes an even more spectacular whorl of about 80 rays joined together into a cap (Fig. 2.1A, on the left); and the single nucleus, 4 cm away from this developmental action, waits until the cap is fully grown before undergoing meiosis on the way to filling the cap with many thousands of gametes. Fig. 2.1B illustrates stages in the simultaneous formation of all the hairs in a vegetative whorl in what is clearly a single developmental event.

My reason for putting the dasyclads with the plants is given in the following quotation from Dumais and Harrison (2000):

> When a definitive identification of the whorl-forming event is reached, will it throw light on the development of anything other than the dasyclads? If the answer to this question is positive, the dasyclads may be very helpful because the patterning of their appendages is one of few instances where an extended fossil record is available to document how a specific morphogenetic process has changed with evolution. Some authors (e.g. Church 1919; Chadefaud 1952; Emberger 1968) not only argued for a natural continuity between algae and higher plants, but also emphasized that the algae account for all the major structural innovations in the plant kingdom. This idea has already been recognized for the Dasycladales. The resemblance between the body plan of *Acetabularia* (rhizoid–stalk–hair–cap) and that of higher plants (root–stem–leaf–flower) had been underlined by Nägeli (1847), Church (1895) and Puiseux-Dao (1962) among others. Solms-Laubach (1895) and Church (1895) drew an even closer parallel between hair and leaf. And more recently, Mandoli (1998) and Nishimura and Mandoli (1992) have identified juvenile and adult developmental phases in *Acetabularia* closely corresponding to such phases in higher plants. Hagemann (1992) and Kaplan (1992) have also discussed general implications of similar development with and without multicellularity.

Dasyclads are the minimalists of morphological development. While, in multicellular plants and animals, so much attention is given to the spatial patterning of on–off switchings of genes, a dasyclad inconspicuously generates equal complexity of form by actions a few centimetres away from its one and only nucleus.

Besides the appearance of a whorl in a single event, simultaneous for all its parts, two other features attracted me to the study of *Acetabularia*. First, the number of hairs in a whorl is highly variable: rarely seen extremes are 3 and 35 hairs – the commonly encountered range is from 7 to 22 hairs – in the vegetative whorls. Variability occurs in successive whorls on the same cell, showing that the number is not genetically fixed. In general, the whorls that attract the most attention in the Dasycladales are the reproductive structures; and indeed, in the whole plant kingdom, these provide a solid basis for classification.

But I became more attracted to making measurements on the vegetative whorls, because of that enormous range of variation in numbers of hairs, which lets one go experimentally after the question: what determines the number, in any individual event?

Second, and perhaps most important, the development at the growing tip is continuously visible, and can be followed day by day or hour by hour in vivo as a study in kinetics. I am a chemical kineticist, and am only really happy when I can study a continuous time sequence. The perils of trying to dig development data out of a set of transmission electron micrographs, each from a different dead and fixed sample – a phenomenon that turns up in animals, plants and protists – are dealt with in Section 7.2.2.

What is there in multicellular plants, the undisputed territory of King Planta unchallenged by King Protista, that offers comparable advantages to the *Acetabularia* whorls for study? To be sure, much of plant development is more accessible to observation than is much of animal development, especially in the viviparous species that many people most want to study. Nevertheless, there are difficulties in trying to follow kinetically even the growth of an apical meristem that is situated in plain air and surrounded by perhaps a pair of leaves that can easily be temporarily pushed aside without damage. But optical microscopy doesn't give very clear views of the cellular array, because the visual contrast in the corrugations of the surface is low. Green and Linstead (1990) devised an elegant method of making a casting of the shape of a meristem, without killing it or inhibiting its growth, by using dental impression materials to make a copy of the surface good enough for the scanning electron microscope to show the cellular ordering.[1]

For any given species of plant, it is usual for a particular kind of branching event at a shoot meristem or a floral meristem to produce always the same number of repeated parts. Numbers of sepals, petals, stamens and carpels in a flower can be used in classification. This in itself is a developmental phenomenon demanding explanation. But again, just as for the unicellular whatever-king-they-are-loyal-to organisms, a highroad into the realm of developmental mechanism should be provided by any case that shows variability in the number of parts. In plant embryos, the angiosperms are as obstinately attached

---

[1] See Kwiatkowska and Dumais (2003) for more recent developments of this technique; also see, for example, Hamant *et al.* (2008) for recent work using green fluorescent protein to image cell walls in live plants (in combination with dynamic modelling).

to particular numbers as in their other branching processes; they make one or two seed leaves and are thus classified as monocotyledons or dicotyledons. In rare cases the number is variable, e.g. 4–11 cotyledons in the angiosperm parasites in the genus *Psitticanthus* (Kuijt 1967). But in the conifers, variability in cotyledon number is quite common, often around an average of about six. The formation of such a pattern is not easy to study in natural development, because plant embryos grow covered by tissues essential for supply of nutrients (the endosperm in the angiosperms, or megagametophyte in the conifers). In recent years, however, means have been found to produce what are known as 'somatic embryos'. From a single seed, by the aid of magic potions known as culture media, embryogenic tissue can be made to grow and to produce embryos perhaps in the tens of thousands. These are, of course, all clones. Two features of these embryos are important. For plant breeders, the possibility of packaging the embryos into 'artificial seeds' with desired genetic characteristics is highly attractive. To students of development with the attitude I present in this account, somatic embryos of conifers have the same advantages as *Acetabularia*. They grow exposed, with nutrients provided from below in a gel in a Petri dish, unobscured by anything like a megagametophyte. Hence they can be observed and measured kinetically. And the number of cotyledons is variable, within a clonal population.

My attention was drawn to these embryos, and to these advantages, by Thurston Lacalli, whose 1973 PhD thesis was my introduction to the field of morphogenesis and pattern formation (Harrison 1993, preface). (By the way, that thesis was upon the desmid pondweed *Micrasterias*, and was in the Department of Zoology, with a zoologist, Alfred B. Acton, as supervisor.) Enquiries aimed at starting a collaboration with a biologist well versed in somatic embryogenesis led to a useful initial interaction with Larry Fowke, but eventually into a blind alley when I found that the work of people aiming at industrial application of these embryos in forestry was tightly wrapped in trade secrets. Also, work so directed usually aims at as much standardization as possible, and therefore tends to go for culture methods that suppress the very variability that I want to study. Here, studies in pure science are not a first step on the road to the obvious application. The applied work is far ahead, but on a different road leading to a different place. (On the relation of pure and applied science, in the topic of thermodynamics, W. J. Moore, 1972, attributed to L. J. Henderson the quotation: 'Science owes more to the steam engine than the steam engine owes to Science.')

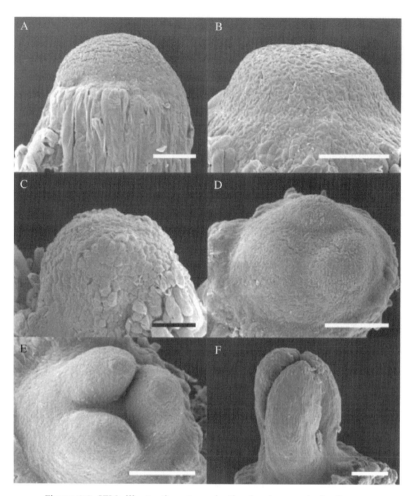

Figure 2.3 SEMs illustrating stages in the development of a three-cotyledon embryo of a hybrid larch; as a sequence for one embryo, this set of pictures is a 'fake' (see text). All scale bars = 250 μm. With permission from Harrison and von Aderkas (2004).

Eventually I found an academic collaborator, Patrick von Aderkas. He supplied me with embryogenic tissue from a hybrid larch, which I was able to observe in continuous development under the dissecting microscope, and make measurements upon. Better pictures, however, are obtained from scanning electron micrographs of fixed specimens. Fig. 2.3 is therefore a fake as a representation of the developmental sequence in one organism. Each picture is from a different dead embryo. But they are chosen to give a good idea of the sequence as

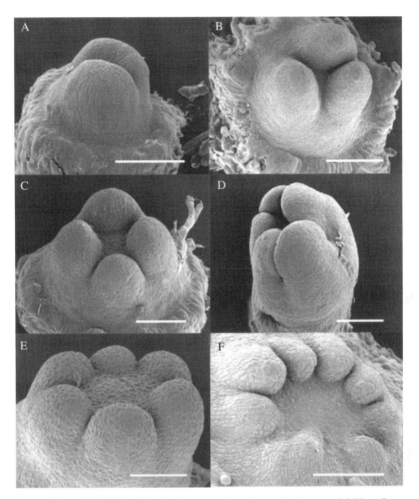

Figure 2.4 SEMs of developing somatic embryos to show variability of
cotyledon number in a hybrid larch: examples with 2, 3, 4, 5, 6 and
8 cotyledons. Scale bars = 250 μm (A–C), 500 μm (D), and 200 μm (E and F).
With permission from Harrison and von Aderkas (2004).

I actually saw it from day to day, for an embryo that made three
cotyledons. Fig 2.3B shows the flattening of the top of the embryo to
an approximate drumhead or cymbal form; Figs 2.3C and 2.3D show the
beginning of cotyledon formation. Fig. 2.4 shows the variability in the
number of cotyledons, with examples from two to eight. In Chapter 3,
I discuss quantitative aspects of this shaping, and show that there is
significant similarity between this phenomenon and the formation of
vegetative whorls by *Acetabularia*.

If indeed there is a significant 'natural continuity' between unicellular and multicellular plants, wherein does that continuity lie, given that the establishment of multicellularity was surely a major evolutionary discontinuity in the structural makeup of organisms? To many developmental biologists, that discontinuity is all-important to the architecture of complex shape in organisms. The generation of that architecture is identified with different switchings on and off of genes in different spatial regions, leading to diverse differentiation states of cells. But *Acetabularia* grows in its warm seas, elongating a main stalk and sending out from it whorl after whorl of hairs, and making all that complexity of shape increasingly distant (in centimetres) from its single nucleus in a single state of differentiation.

2.2 ENTITIES AND WAVES

I believe (or preconceive) that the 'natural continuity' is to be sought in the general nature of developmental mechanisms. These may act most often to control spatial patterning of cell differentiation states. But they can pattern quite other things that control growth rates of parts of a single cell surface, or spatial distributions of mechanical stresses, or spatial patterns of deformability in response to stresses. Here, there is a feature that overrides even the difference between Green's 'biophysical' and my 'physicochemical' view of the most promising kinds of mechanism. When a pattern such as a set of whorled structures first starts to arise, it may do so as a spatial entity, a set of hills and valleys on a surface, or as a chemical entity, regions of concentration and dilution. But in either case, if the pattern is a single entity, these things should appear first as harmonic waves appropriate to the overall geometry, e.g. the instantaneous shape of a vibrating cymbal such as the top of Fig. 2.3B. In this book, the terms 'harmonic' and 'wave' usually do not imply anything moving in an oscillatory manner in time, but rather a shape such as would be seen in an instantaneous snapshot of a vibrating object. This may be a shape in one, two or three spatial dimensions. For 1-D, the object could be a violin string, sounding in a pure tone so that at any instant its shape is a simple sine wave. For 2-D, the object could be a circular disc; for 3-D, the rather more rarely encountered hemispherical bell. This last, being a thin shell rather than a solid hemisphere, has of course both 2-D and 3-D aspects. In most of biological development, the shapes of 3-D organisms are achieved by actions in thin layers. Fully 3-D events are rare.

I mention the disc and the hemispherical shell because the morphogenetic regions of plants often approximate one or other of those circularly symmetric shapes. The active regions commonly cap a stalk of circular cross-section, and most morphogenetic events are in the class of symmetry-breaking events, in which the circularity is lost. The harmonic waveforms appropriate to the drumskin and the hemispherical bell are very important in the study of plant morphogenesis. For a drumskin attached to a rigid circular border, these are a combination of circular up-and-down ripples with alternating up-and-down pie-slice patterns. The ripples are no longer a simple sine wave function, but are a less familiar function to most people, known as a Bessel function (or, translating the German term for them, cylinder function). Roughly, these are sine waves modified by a function of the radius. With the aid of tabulated and plotted values they can be used without much need to plunge into the mathematics. I illustrate this in Chapter 3, showing a quantitative fit between the larch data and harmonics on a disc.

Meristems or growing tips do not always flatten before undergoing a symmetry-breaking event, and the hemispherical shell harmonics also need to be considered. Some of these are actually very familiar shapes to all chemists and indeed have been encountered by most scientists. They are the angular components of the hydrogen atom orbitals, of which the shapes of s, p and d orbitals are the most familiar. Curiously, to consider the kind of harmonic that would lead to dichotomous branching of a growing tip, one has to go further and use the harmonic that corresponds to an f orbital, something that most chemists never have to consider. Also, at least in reaction–diffusion theory of how these harmonic patterns may arise, it turns out that the pattern for dichotomous branching competes on an equal basis with another one of the 'f orbital' type that is annular, and would account nicely for flattening of a dome-shaped meristem or growing tip (as from Fig. 2.3A to Fig. 2.3B) without any loss of circular symmetry (Harrison *et al.* 2001).

It seems that this geometrically simple and probably most ancient form of branching in plants is not at all simple to account for mechanistically. Again, just as for whorl formation, I think of multicellular and unicellular examples together. The desmids of genus *Micrasterias* make their stellate shapes by repeated dichotomous branchings (stages summarized in Fig. 2.2C). In my group, the capabilities of reaction–diffusion for controlling such a sequence have been studied extensively by 2-D computations to the extent of simulating the morphogenesis of very diversely shaped species within the genus (Harrison and Kolář 1988;

Holloway and Harrison 1999a; and this book, Chapter 5). But there are obvious questions that cannot be answered in 2-D work. Most particularly, how does the organism manage to control successive branchings so that they are all in the same plane? At the present state of our 3-D work, it is evident that there is a strong tendency for a second dichotomous branch to be in a plane at right-angles to the first (Harrison *et al.* 2001).[2]

Yet another kingdom has claims to this phenomenon. The fungus *Neurospora crassa* sends out tip-growing hyphae that branch, usually by appearance of single lateral branches from the main stem of the hypha in apparently random positions. But there are mutants in which this kind of branching is suppressed in favour of dichotomous branching at the growing tip (Virag 1999). This example offers some hope that the 'top-down' macroscopic-scale approach of pattern-formation theorists and the 'bottom-up' molecular-scale approach of geneticists may be closer to meeting in the middle than has usually seemed to be the case in the past three decades. (It is, however, quite likely that the dichotomous branching mechanism in *Neurospora* mutants is somewhat different from the branching mechanisms in the tip-growing algae, because of a huge difference in spatial scale. Fungal hyphae are of order $1\,\mu m$ in diameter, while (Fig. 2.5) *Acetabularia* growing tips are of order 100 times wider. This latter dimension places the spacing of new organs arising on an *Acetabularia* growing tip in the same order of magnitude as such events in very many higher and lower plants and animals. This quantitative aspect could possibly be a more important criterion for probable similarity of mechanism than the matter of which kingdoms two organisms being compared are in, or whether each is unicellular or multicellular. (My interest in the little seaweed has always been that its invisible Organizer takes a minimalist attitude, saying to us: 'Hey, look, I can do it without having an array of nuclei that can have different on–off switchings of genes in different parts of the patterned region.')

2.3 NUMBERS OF PARTS, SPACINGS, WAVELENGTHS

For a patterning phenomenon showing variability in the number of repeats of an organ produced in a single patterning event, my first questions are: what determines the number? What is constant, and

---

[2] See Holloway and Harrison (2008) for mechanisms that make successive branches coplanar.

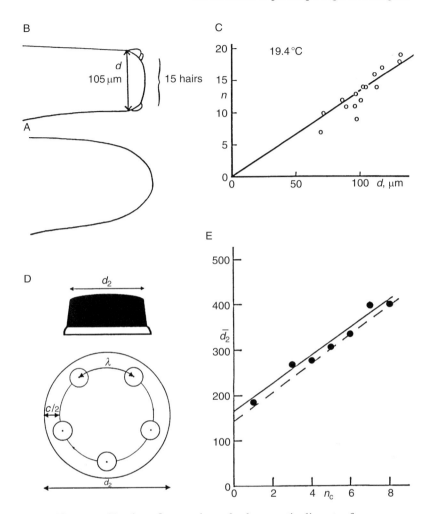

Figure 2.5 Number of organs in a whorl versus tip diameter for: A–C, unicellular *Acetabularia acetabulum*; D–E, multicellular *Larix × leptoeuropaea*. A: tip shape during continuing tip growth without whorl formation; B: flattened tip with hair primordia at usual stage for measurement of diameter d; C: number n of hairs in whorl versus diameter d. D: side view of stage for measurement of diameter d_2 and top view showing d_2, inset $c/2$ and spacing λ. E: diameter versus number of cotyledons: circles and solid line, experimental data; broken line, fit to Bessel functions (Fig. 3.5). With permission: A–C from Harrison *et al.* (1981); D and E from Harrison and von Aderkas (2004).

what is varying, in the patterning region at the time when the pattern is determined? Fig. 2.5 illustrates the experimental pursuit of this question for both *Acetabularia* vegetative whorls (A, B and C) and *Larix* somatic embryos (D and E). In both cases, it turned out that the flattening of the growing tip or apical meristem provided a state in which a diameter (d or d_2) could be measured at a morphologically well-defined position that looks to be closely related to the boundary of the patterning region. In both cases, this diameter varied greatly from one specimen to another. So did the number of organs in the whorl. But a plot of diameter versus number showed something more precisely controlled: a linear relationship, the slope of which indicates a constant spacing between cotyledon primordia in *Larix* (E), or 1/spacing the way round the *Acetabularia* data are plotted in C. The line in C goes through the origin, indicating a spacing effect round the outside edge of the circle of diameter d. The line in E does not go through the origin, indicating a spacing λ controlled around a smaller circle on the flat top of the embryo. Actually, C is the first and only plot we published of this functional relationship for *Acetabularia* (Harrison *et al.* 1981). Later work on many more specimens showed that the line goes closest to the origin if we take the primordia as circles packed just inside the outer edge at diameter d.

What is to be explained, therefore, by any theory of the pattern formation in both of these cases is a phenomenon of constant spacing between adjacent primordia as they first form. My a-priori choice of theory to pursue is Turing's (1952) reaction–diffusion theory. It indeed explains patterning in terms of spatial waves of concentration of chemical substances, with wavelengths determined by rates of reaction and of diffusion that should be constant under controlled experimental conditions (temperature, culture medium composition, etc.). So the theory stands up to the first test against experiment.

In the consideration by biologists of whether the reaction–diffusion concept could have any relevance to the developmental phenomena they observe, there has been some confusion about the constant-wavelength concept. This arises because, in phenomena where a constant number of parts must be produced (e.g. insect segmentation), there is commonly a variation in length between individuals in a population of embryos, but constancy in the pattern generated. This has appeared to be a stroke against reaction–diffusion theory. The objection was best put by Waddington (1956) in the words: 'In patterns which behave in this way, the distance apart of the various elements cannot be fixed by any chemical wavelength dependent on the

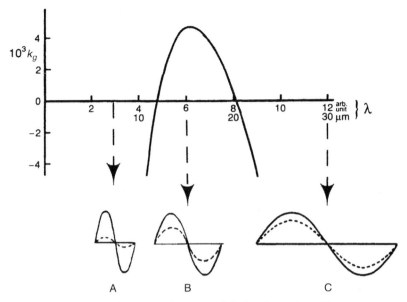

Figure 2.6 The band-pass character of Turing dynamics with respect to pattern wavelength (see text). k_g is exponential growth rate of pattern amplitude at wavelength λ. A, B and C are concentration versus distance plots for a simple chemical pattern of one wavelength. Solid curves are concentration of an activator, broken curves concentration of an inhibitor. Pattern A decays because of high concentration gradients giving fast diffusion; C decays because of high inhibitor/activator ratio. B has the balance giving growth.

unchanging values of rate constants, diffusion constants, etc. Rather the pattern must arise as a whole within the boundaries of the material available.'

This objection is invalid. The Turing theory does not indicate a precise wavelength that a growing pattern must have, but has a 'band-pass' character, such that very short wavelength patterns will be suppressed, so will very long wavelength ones, but there is a range of wavelengths of pattern that can grow. Suppose, for instance, that we consider *Acetabularia* hair patterning to be essentially 1-D, around a circle that I have called the 'A pattern' in Figs 2.1C and 3.1. Suppose further that a particular whorl has ten hairs. This simply means that a pattern of the wavelength defined by 1/10 of the perimeter grows faster than one of the wavelength defined by 1/9 or 1/11; and the same number of hairs, ten, is likely to be established over a range of about 10% in growing-tip diameter d. Fig. 2.6 illustrates the band-pass character of Turing dynamics by a plot of the exponential growth rate k_g of the amplitude of a sine wave pattern versus its wavelength. Short and long wavelength patterns A and C would decay (negative k_g), while pattern B, of intermediate wavelength, would grow. But the wavelength of this pattern, shown as six arbitrary units, could be anything between 4.8 and 8, the range of wavelength with positive growth. (I return to this topic in Chapter 6.)

The problem here has been that applied mathematicians, physicists and physical chemists are commonly acquainted with partial differential equations (ones with time and space variables) and know that each of these usually has a multitude of possible solutions of which one is selected by what the boundary conditions happen to be, but not in the rigid way that the quotation from Waddington suggests. For simpler patterns with only two parts, such as the patterning of two types of cell in the 'slug' stage of the slime mould *Dictyostelium discoideum*, Lacalli and Harrison (1978a) showed that a Turing mechanism could generate the same type of pattern over a range of about a factor of five in the length of the slug. So constant-spacing effects always involve some statistical range in the spacing. It gets to be a smaller range, on a % basis, as the number of parts in the pattern becomes larger, which again makes *Acetabularia* an ideal organism in which to find the effect.

The most fundamental point about a constant-spacing effect, however, is not to do with the complexity of a pattern, but with the ability of chemistry and physics to set up quantitative scales of distance. The measurement of distance and the ability to set up arrays of repeated parts are two sides of the same coin. We measure distances by first making a ruler marked out in repeated parts. How does a physical or chemical mechanism determine the size of a unit of distance? For chemistry, let us start with rates of reaction and diffusion in the most general terms. A diffusivity D has dimensions of $(distance)^2 (time)^{-1}$, e.g. $cm^2 s^{-1}$. The simplest kind of chemical rate constant k is in $(time)^{-1}$, e.g. s^{-1}. If we combine these two in the form $(D/k)^{1/2}$, we have a quantity with dimensions of distance, e.g. units of centimetres. I have previously pointed out (Harrison 1993, p. 36), with an example of a simple monotonic gradient of concentration, not a wave pattern, that all quantitative measures of distance in reaction–diffusion theory are more or less complicated variations on this theme. Consider these equations for a wavelength λ:

$$\lambda = 2\pi(D/k)^{1/2} \tag{2.1a}$$

$$\lambda = 2\pi \left[D_x D_y / (-k_2 k_3) \right]^{1/4} \tag{2.1b}$$

$$\lambda = 2\pi (D/k)^{1/4} \tag{2.1c}$$

The first is the generic form, and is very useful for playing with likely values of diffusivities and rate constants to see what kinds of wavelength they give. The second is an approximate wavelength expression for 'two-morphogen' or 'activator–inhibitor' reaction–diffusion, with diffusivities for two substances, X and Y, and two rate constants that I am not about to attempt to explain at this point. (Note the minus sign; one of the rate constants is negative, which should warn the reader of complications.) But Eq. (2.1c) is not to do with reaction–diffusion at all. The use of symbols D and k in it almost looks like a piece of amusing trickery, for they do not represent a diffusivity and a rate constant. The equation is from 'The mathematics of plate bending' (Rennich and Green 1997) and belongs to the mechanical buckling theory of formation of wavelike patterns. Here, D is the flexural rigidity of a plate or beam, and k is the force constant (coefficient of elasticity) of a set of springs that is holding the plate or beam down as it tries to bend upwards and downwards into a wave shape (see Section 4.2.) There can be some remarkably close correspondences between some aspects of theories that are otherwise radically different in the ways they explain pattern formation, and these can often be found in the expressions they lead to for wavelengths (pattern spacings, quantitative scales of distance). (Eq. (2.1) refers only to theories with chemical dynamics alone or mechanics alone. I mention mechanochemical theory in Section 4.2, but do not give an extended account of it in this book.)

Now that we have some quantitative expressions, let us consider again the 100-fold contrast in linear dimensions between growing tips of single cells in the fungus *Neurospora* and the alga *Acetabularia*. Biological patterning is unkind to trendy physical concepts of self-similarity and scale invariance. Particular mechanisms tend to demand particular time and distance scales, and it is remarkable how many and varied patterning phenomena in living organisms choose tens of micrometres for the latter. The size of a fungal hypha tip is at the low end of the reasonable range. Here, I am getting into another contrast between the attitude to pattern initiation I am advocating and more conventional views in biology; and again the contrast is well put in Green *et al.* (1996):

> Most models for plant pattern are based on patterned controls and assume
> that the organ initiating event is momentary. It involves the
> instantaneous production of a new organ as a point or as a well-defined
> district (Douady and Couder 1996) . . . We argue here that, instead
> of the critical event initiating periodicity being an 'abrupt activation
> at a point', it could equally well be 'gradual undulation of an area'.

The time and space aspects raised here are: how gradual, and what area? For the reaction–diffusion possibility, we may appeal to the most general form of the relationship, expressed in Eq. (2.1a). It contains a rate constant k, which has something to do with the word gradual, and can be seen that way if we replace it by its reciprocal $1/k = \tau$, a time:

$$\lambda = 2\pi(D/k)^{1/2} = 2\pi(D\tau)^{1/2}, \text{ leading to } D = \lambda^2/4\pi^2\tau \tag{2.2}$$

My estimates of τ are usually based on what one sees in following a developmental event through the morphological emergence of pattern. Such observations cannot eliminate the possibility that an organ-initiating event is 'momentary', 'instantaneous' or 'abrupt'; but to my mind the better guess is that pattern formation has continuity through a time period up to the first sign of morphological expression, giving a value of a few minutes for τ. For an order-of-magnitude illustration that works out in round figures, I shall take $\tau \approx 250\,\mathrm{s}$ and $4\pi^2 \approx 40$, so that Eq. (2.2) becomes $D \approx \lambda^2/10^4$, whence:

$\lambda(\mu\mathrm{m})$	0.1	1	10	1000
$D(\mathrm{cm}^2\mathrm{s}^{-1})$	10^{-14}	10^{-12}	10^{-10}	10^{-8}

The range of spacings from 1 to 1000 μm corresponds to a tolerable range of diffusivities for processes between cells, within cells or upon cell surfaces. It corresponds also to the working range of the light microscope, which therefore is the instrument of choice for people studying developmental patterns as entities. It is fortunate that this instrument permits in vivo studies over a continuous time period.

3

Whorled structures

In this chapter, I describe detailed experimental data relevant to mechanisms of pattern formation for two biological systems that I have studied. For the *Acetabularia* whorls, I believe the evidence points strongly to a reaction–diffusion mechanism; for the *Larix* cotyledons, it indicates strongly that the mechanism is waveforming and hence that the embryo apex is behaving as a coherent whole during pattern formation. However, the evidence does not distinguish between reaction–diffusion and mechanical buckling. I reserve some account of the latter until the following chapter (Section 4.2); the mechanism is, however, equally applicable to dichotomous branching and whorls of a larger number of parts.

3.1 *ACETABULARIA ACETABULUM* VEGETATIVE WHORLS

3.1.1 The morphological event

Trying to cross disciplinary boundaries, one is beset with many perils. Some of these are rather trivial things to do with word usage. Having read often of 'whorls of stamens' and suchlike, I had assumed without the benefit of a dictionary that 'whorl' meant a set of structures arranged in a ring. I have recently been warned that to some biologists, the whorl is only the ring itself. I have been calling that an annulus, and the set of structures disposed around it the whorl. In this book, I shall adhere to that usage, deferring to the other one only in the title of this chapter. (Much worse happens, and if interdisciplinary work is to thrive, scientists need to be multilingual in their terminologies. In my previous book I frequently used the term 'optical resolution' in the chemist's sense of making a substance with an asymmetric molecule so that it contained only, say, the left-handed form. Only long after publication did I discover that when you say 'optical resolution', a physicist thinks only of how small a thing you can see under a microscope.)

But crossing boundaries becomes much more difficult when it involves rapid and fluent comprehension of a new set of principles, whether these are ones that have to be expressed in mathematical terms or, on the other hand, by using the panoply of terminology of molecular genetics. Biologists often have trouble with the former; I, with the latter. And I have known a mathematician to get interested in phyllotaxis (disposition of leaves on a stem) and to become terror-stricken on encountering the word 'plastochrone'.

My problem in getting into any depth in describing particular examples of research on pattern formation is how and where to say how much or how little about such concepts as the formation of pattern by reaction–diffusion. First, to get the real feel of how such a mechanism works, one has to live with the equations for months. Second, and perhaps largely as a result of that first point, there is a fair amount of a-priori hostility among biologists to recognizing any promise in this kind of theory for the explanation of events in biological development. This means that if I feature this theory too much, too early, I may create an obstacle to many people continuing to read the book, which is not intended as a polemic upon that theory. In *Acetabularia* whorls there is, to my mind, significant evidence pointing to a reaction–diffusion mechanism; but there are more general things to be said. We are now ready to consider Fig. 3.1, which is the same as Fig. 2.1C, but with the reaction–diffusion taken out of the middle of it. (To anyone whose experience of tip growth is with fungal hyphae of diameter about 1 μm, I should mention that the tip of an *Acetabularia* main stalk has diameter of order 100 μm, and is more likely to have comparable features to multicellular plants than to fungal hyphae, which I get to discussing in Chapter 4.)

The sketch at the lower left represents an *Acetabularia* growing tip when it is doing nothing more than elongating the main stalk of the cell. Shading on the tip represents a graded rate of growth, highest at the extremity or 'pole' of the tip and much lower at the equator; or, the shading can represent the graded concentration of some growth catalyst (designated substance A) that maintains that grading of the rate. The solid line at the equator raises an important question: as the stalk elong-ates by addition of cell surface materials at the growing tip, how does the growth activity remain confined to a dome on the end of the stalk? The equatorial boundary must continually move up with the dome. If it

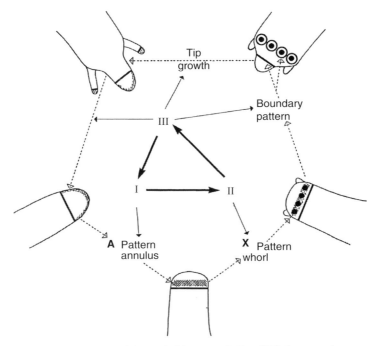

Figure 3.1 Schematic of the probable events (I, II and III) that must happen for an *Acetabularia* hair whorl to form and for each hair to develop as an independent tip-growing structure.

did not, but remained stationary, the equator would become the open end of a pipe on which the growth activity would 'blow a bubble'. (That kind of growth also happens in the Dasycladales; see the pictures of species in the family Dasycladaceae in Berger and Kaever 1992, and developmental sequences in Dumais and Harrison 2000.)

This kind of tip growth can continue for a few days; but from time to time, the tip flattens (Fig. 3.1, bottom centre; we are going anticlockwise). The shading suggests that the growth rate and region of high concentration of substance A have shifted to a band near to the equator, called the annulus, to bring about the flattening. Whenever this happens, the beginnings of a whorl of hairs appear within a few hours (lower right in Fig. 3.1). The black dots and designation 'X pattern' suggest that another catalyst, X, has become patterned in a spatially periodic way around the annulus, to give rise to growth rate maxima likewise patterned. That, at least, is the way that I first thought of it when I started observing the development in vivo and making measurements to study what determines the number of hairs. A closer look at the details of the developing morphology provides a different story.

The locations where hairs are about to emerge are marked first by formation of pits on the inside of the cell wall. From Dumais and Harrison (2000):

> Werz (1965) published evidence for the nature of the lysis process. First, uncharacterized proteins accumulate in the morphogenetic region and indicate the future location of the appendages. Second, a close association between the endoplasmic reticulum and the lysis sites is seen, suggesting that local secretion of hydrolytic enzymes is responsible for the wall lysis. . . . The spacing and tip diameter increase during further development of the whorl but the number of lysis sites, which eventually determines the number of lateral appendages, remains fixed.

How does this view of the probable chemistry affect my pursuit of mechanisms in which the whorl is first specified by patterning of a 'substance X'? Not at all, actually. X may be one of the substances mentioned in the above quotation, and not directly a growth catalyst. But, more indirectly, it has to lead to growth of the hairs. For computations on the interaction of chemical patterning and growth, it remains valid to abbreviate the chemistry at each lysis-to-growth site by a single substance X. More is said about that kind of computation in relation to *Micrasterias* in Chapter 5.

Each hair, starting from a small bulge in the cell surface, elongates into a narrow stalk. This needs a repeat, for each hair, of a growing tip as already discussed for the main stalk. And that requires drawing of a new boundary for each tip, as shown by the open circles in the sketch at the top right of Fig. 3.1; and likewise for the resumption of growth of the main stalk, with its tip boundary shown as a solid line. (For completeness: each hair also undergoes secondary branching, but I am not discussing that here.)

Without as yet committing myself to any particular mechanism of pattern formation, I have now identified three processes demanding explanation, I, II and III in Fig. 3.1. When I pursue chemical-kinetic reaction–diffusion theory as a possibility, I call these processes 'feedback loops I, II and III' because they all require that dynamic character to be pattern-forming. That takes us back to the insert in Fig. 2.1C of some putative chemistry involving substances called S, A, B, D, E, X, Y_1 and Y_2. How much does one need to know about the dynamics that these things are allegedly involved in to follow the strategy I used in studying *Acetabularia acetabulum*?

There is a conspicuous contrast between my earlier studies on this example of morphogenesis (Harrison *et al.* 1981, 1988, 1997;

Harrison and Hillier 1985) and the details of lysis pits and so forth to be found in Dumais and Harrison (2000). The former studies were necessarily in vivo, designed to follow a very large number of specimens and try to pick up a stage as near as possible to the moment of whorl pattern formation and make spatial measurements on it. The latter were on dead preparations of cell wall 'ghosts'. The kinds of information obtained by each technique would not be readily available from the other. Fortunately, they turn out to be complementary, because the hair primordia seen in the in vivo work are in equal number to the lysis pits and appear very soon afterwards, when there has been very little change in size.

3.1.2 Reaction–diffusion: looking for circumstantial evidence with a magnifier

In Section 1.4 I discussed an approach to complex reaction mechanisms, referring to examples that have the properties of explosion and of oscillatory reaction. The point of both of these examples is that one may have a vast amount of detail on some sequence or network of steps in a reaction mechanism. But to find therein why it behaves in some known way, or alternatively to predict some unknown property that should arise from the mechanism, one needs to have in mind some motif, involving very few steps, that confers the property of interest on the reaction dynamics. The essence of the two strongly contrasted ways of approaching mechanisms of symmetry-breaking events in biological development is:

(1) The molecular-genetic approach is to collect details until one has a schematic of the whole reaction network and then examine its dynamic properties. Only in quite recent years has it come to be generally realized that people with mathematical expertise may be of use at that stage, because it is only quite recently that fairly complete networks of gene control have started to become available. The approach of amassing details of genes, proteins, sugars, etc. is currently being expressed by a proliferation of words ending in '–omics'. I have recently encountered: genomics, proteomics, 'signalling/transcription-omics', degradomics, metabolomics and glycomics. (The word genome came, around 1930, from a fusion of gene with chromosome. Somehow, the Greek soma (body) mutated so as to lose its initial letter, and then hybridized promiscuously with whatever word happened along.)

(2) What I shall call the 'reaction–diffusion' approach is to try to establish the presence of the motif that confers on the network the symmetry-breaking or pattern-forming property (I take these terms as essentially synonymous) so that, later, it will be recognizable as useful to hunt for that motif in a network schematic. This approach involves doing kinetics of the generalized 'A + B → D + E' type, where the biochemical identities of the substances are not yet known. Many people, especially inorganic and organic and biological chemists, feel uncomfortable with that aspect and criticize physical chemists for using it. But once, when teaching a kinetics course for the first time with the substantial aid of two complete sets of notes, one from an organometallic chemist and the other from a physical chemist, I discovered that it was the former who used 'A + B → D + E' and the latter who did it all with specific examples. Chacun à son goût. And I wish that the value of that little French maxim were more widely appreciated among scientists facing apparently intractable problems. Diversity of approaches should then be the essence of the broad strategy. (The astronomer Fred Hoyle once wrote: 'When a problem remains unsolved, general opinion must be wrong.' Notwithstanding that most astronomers today do not agree with Hoyle's concept of the origin of the universe, I think he was right in that statement.)

Some motifs of reaction–diffusion dynamics are discussed in Chapter 5. They include:

(1) A tendency for repeated structures, under fixed conditions of temperature, culture medium composition, etc., to have a fixed spacing between them, regardless of the size of the region they are filling (with the qualifications on this statement already made at the end of Chapter 2, to cope with too rigid an idea of the 'fixed chemical wavelength'). This is most noticeable if the number of structures is large, and the size of the patterned region is very variable. *Acetabularia* vegetative whorls are ideal for this test, and *Larix* somatic embryo cotyledons have also proved very informative (Fig. 2.5).

(2) Again, most noticeably for large numbers of repeated structures, dependence of the spacing between them on: (a) temperature; and (b) concentrations of components

of the culture medium, especially ones that might be
the inputs A and B or substances involved in generating
those.

In thus approaching the possibility that a pattern is formed by
chemical dynamics, the most important generalization is that one has
to think of a spacing between repeated parts of a pattern as being not
something structural and static, but rather a kinetic parameter, with
the kinds of dependences on temperature and concentrations that one
expects of rates of reaction. The results that I and my group obtained for
Acetabularia whorls at once look quite upside-down. We found that the
spacing between hairs decreased as the temperature was raised (Harrison
et al. 1981). Having discovered that the spacing was dependent on calcium
concentration in the culture medium, we established that the spacing
went down as the calcium concentration increased (Fig. 3.2A). Chemical
reactions usually go faster if temperature or reactant concentrations
are raised.

The results we found are, however, actually very compatible with
reaction–diffusion theory. As discussed in Chapter 5, the opposition
between diffusion tending to make spacings longer and reaction rates
tending to shorten them leads to expressions for spacings in which the
reaction rates appear as reciprocals. This gives temperature and concentration dependences in the observed directions. In particular, we were
able to follow up the calcium concentration dependence quite extensively
(Harrison and Hillier 1985; Harrison *et al.* 1988, 1997) and eventually
arrived at a suggestion for how calcium plays a role in the pattern-forming
mechanism, which is summarized diagrammatically in Fig. 3.2B.
Between Figs 3.2A and 3.2B there is a lot of work by a number of people.
Some of the results are presented as a lot of straight lines in Fig. 3.3.

And at this point in the writing I got stuck and remained stuck
until I realized that I should just say that that had happened and
explain why, because in fact this touches on the philosophy of trying
to make some kind of scientific advance by using 'circumstantial evidence'. To my mind, that sort of evidence pointing to a reaction–diffusion
mechanism for this phenomenon is very strong. But the only kind of
jury that is likely to convict this hapless little seaweed on a charge of
harbouring reaction–diffusion (a serious offence to many biologists) is a
jury well acquainted with how one goes about analysing biochemical
kinetics, especially the kinetics of enzyme-catalysed reactions. This book
is intended to be addressed to a very general readership of all kinds of
people with some scientific education, not just to those who will at

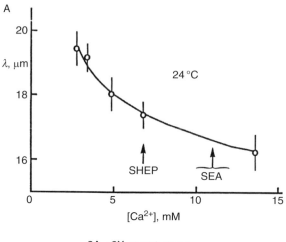

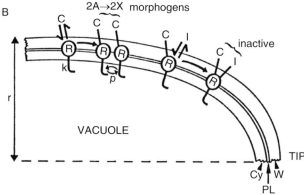

Figure 3.2 A: looking for spatially quantitative dependence of a pattern-forming event on a chemical concentration. λ is the spacing between adjacent hairs in an *Acetabularia* hair whorl. The concentration of calcium, $[Ca^{++}]$, is in Shepard's culture medium, a form of artificial seawater. SHEP indicates the calcium level usually put into that medium; SEA indicates the commonest value in natural seawater. B: schematic of a putative mechanism for the calcium effect, based on much physicochemical data (as shown in Fig. 3.3). The sketch shows a section through one side of an *Acetabularia* growing tip. Cy: cytoplasm; PL: plasmalemma (cell membrane); W: cell wall; C: a calcium ion; R: a receptor protein molecule for calcium, which is activated for kinase activity (catalysis of phosphorylation) by the calcium binding; k: the piece of R inside the cell that has the kinase activity; p: k regions of two R cells each busy phosphorylating the other; I: an inhibitor of this activity of R, being in this work the substance EGTA (Fig. 3.3C).

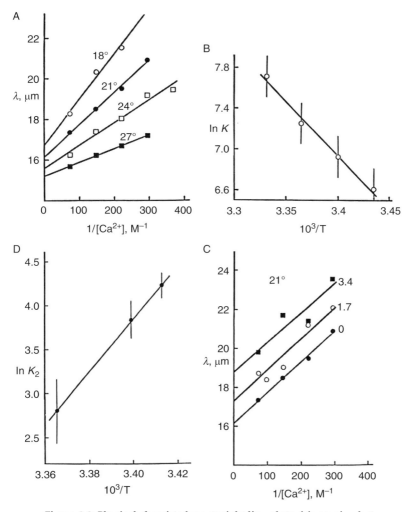

Figure 3.3 Physical chemists love straight-line plots. λ is spacing between adjacent hairs in an *Acetabularia* whorl. K is the equilibrium constant for binding of Ca^{++} to a putative receptor R (Fig. 3.2B). K_2 is the binding constant for EGTA (inhibitor I in Fig. 3.2B) to RCa^{++} (RC in Fig. 3.2B). A: plots of spacing versus reciprocal of calcium concentration; the 24° line is from the same data plotted in Fig. 3.2A. These plots are analogous to enzyme kinetics (Michaelis–Menten) if the spacing λ depends inversely on a reaction rate constant. B: binding constants calculated from plots in A versus reciprocal of temperature, i.e. van't Hoff plot, from which enthalpy and entropy of binding can be calculated. C: plots like those in A for culture media containing CaEGTA at concentrations 0 (same plot as 21° plot in A), 1.7 and 3.4 mM. These plots show the expected behaviour for 'uncompetitive inhibition'. D: van't Hoff plot for the inhibitor binding to RCa^{++}.

once understand if I say that Fig. 3.3A is analogous to a Lineweaver–Burk plot of Michaelis–Menten kinetics.

Quite apart from such terminology, I am by no means sure that all kinds of scientist will become as excited as I am when the data turn out to be expressible as straight lines on a graph. This curious fetish for straight lines is often regarded as a manifestation only of the psychopathology of physical chemists. But I am one of those, and I have got excited about the set of linear relationships assembled in Fig. 3.3, concerning the effect of extracellular calcium on the spacing λ between adjacent hairs of an *Acetabularia* whorl, as defined in Fig. 2.5.

Fig. 3.3A shows how the spacing λ changes with the *reciprocal* of calcium concentration. And at an early stage, when all I had was four of the five data points in Fig. 3.2A, I got quite ecstatic to see that the reciprocal plot turned the curve into a straight line, and even gave a talk at a meeting on that bit of data only. The diversity of attitudes of different groups of scientists to various kinds of data is quite remarkable:

(1) When I gave a talk on Turing reaction–diffusion theory to my home chemistry department, one of my colleagues remarked 'It's nice to see that such an apparently difficult problem can be explained by a couple of differential equations.' I thought 'Oh, dear, I haven't got over to him that most biologists reject this theory.'

(2) A biologist said to me 'Wouldn't it be a pity if the explanation turned out to be just physical chemistry', wrinkling his nose as if experiencing a bad smell.

(3) After a talk in a chemistry department, one of the audience said to me 'What I like about your stuff is that you go straight from the green stuff to the differential equations.'

(4) At a poster show at a developmental biology meeting, my collaborator David Holloway was waylaid by a posse from another modelling group who started their work from molecularly known genetic networks, saying 'Why don't you guys want to use real molecules?'

Let us return to Fig. 3.3, starting at A. First, the range of values of λ is of some interest. The spacings are in tens of micrometres, that is, hundredths of a millimetre. This order of magnitude is very common for the spacing-out of repeated parts when they first appear, not only in unicellular plants but throughout life. The stripes of gene expression that mark the beginning of segmentation in a fruit fly embryo and the

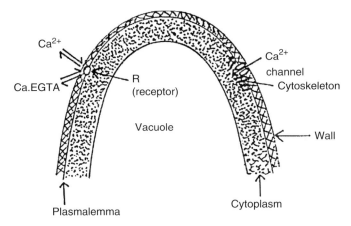

Figure 3.4 Contrast of cell surface (L) versus cytoskeletal (R) possibilities for the mechanism of the effect of extracellular calcium on spacing λ in *Acetabularia* whorls.

somites that mark the beginning of backbone and nervous system segmentation in a human embryo have similar spacings. Such spacings are observable and measurable at fairly low powers of magnification.

Second, calcium concentrations are in the millimolar range (natural seawater is about 10 mM in calcium). These are enormous values for anything interacting with a living cell. Calcium is an essential element for many of the workings inside cells, but its intracellular concentration is very carefully controlled, and too much of it is poisonous. These extracellular values are greater by a factor of order 10 000 to 100 000 times greater than intracellular calcium concentrations. The little seaweed must be expending appreciable effort protecting itself against this hostile exterior. This suggests that the calcium concentration inside the cell is very little affected by changes outside. Amtmann *et al.* (1992) showed with a calcium-specific electrode that cytosolic free calcium in *Acetabularia* is at 560 nM, and that complete removal and re-addition of 10 mM external calcium changed the value inside by only 50 nM.

This has a bearing on where one should look, in the architecture of the organism, for the mechanism for λ determination and variation that Fig. 3.3 gives information upon. There are two obvious possibilities (Fig. 3.4), associated with two rival published theories (Harrison and Hillier 1985; Goodwin and Trainor 1985; Brière and Goodwin 1988; Harrison *et al.* 1988, 1997). The Goodwin concept is that extracellular calcium enters the cell and controls action of the 'cortical' cytoskeleton immediately inside the cell that produces the whorl patterning (Fig. 3.4, on the right). The Harrison concept is that calcium binds to the extracellular side of an integral membrane protein (R, for receptor of calcium), thereby activating R so that it can act as a precursor A for an activator morphogen X through phosphorylation activity on the intracellular side (Fig. 3.4, on the left, and Fig. 3.2B). Both concepts draw upon well-established features of cell biology: Goodwin's upon the mechanical properties of the cytoskeleton; mine upon the existence of integral membrane proteins with external receptor sites and internal kinase activity. I prefer my idea (what else did you expect?) because my evidence (those straight lines in Fig. 3.3) seems to point to a rather direct effect of extracellular calcium concentration, binding to extracellular receptors and therefore unhindered by interference with its concentration effect by that calcium-excluding cell surface.

So what do those straight lines in Fig. 3.3 tell us about the binding of calcium to an extracellular receptor if we decide to follow up the idea that that is what is happening? Just about everything that can be expressed in numbers to give the strength of binding, how it changes with temperature, and what that means in terms of the thermodynamic properties of the binding (energy and entropy). To see without plunging into all the equations that that kind of information is to be found there, first recognize that my interpretation of what determines λ makes it inversely proportional to the concentration of some substance on the surface that is active as an input to the patterning processes. In Fig. 3.2B, that substance (A) is a combination of a cell surface protein R with calcium C. As already mentioned, the spacing λ is expected to be inversely proportional to the concentration of inputs such as A. This means that the plots in Fig. 3.3A are plots of 1/[bound calcium] versus 1/[free calcium]. (Square brackets, in the common chemical convention, are to be read as 'concentration of'.) In this interpretation, every experimental point in Fig. 3.3A, and indeed every point read off the lines drawn through those points, carries information on how much calcium is bound to the receptors R and how much is not. Further, the left-hand end of each line, because the plotting is for reciprocals of concentrations, represents infinite calcium concentration, at which all the receptors R would be plugged up with bound calcium. It should now be obvious that the graph carries enough information to enable the binding constant, i.e. the strength of binding of calcium to R, to be found at each of the four temperatures for which data are plotted. Readers familiar with this kind of analysis are immediately going to get suspicious. It is obvious that there is more calcium bound at 27 °C than at 18 °C, and that means that energy is absorbed in the binding process, the opposite of what one usually expects. In fact, calcium often takes part in such endothermic binding, which has to do with the positively charged ion often binding to negatively charged groups to give a neutral complex that releases water from the constraints imposed by electric charges. (For such readers: binding constant $744\,M^{-1}$ at 18 °C and $2240\,M^{-1}$ at 27 °C; $\Delta H^{\circ}_{298} = +88$ kJ mol^{-1}, $\Delta S^{\circ}_{298} = 356$ J mol^{-1} K^{-1}.) Fig. 3.3B is a plot of the temperature dependence of the binding constant from the lines in Fig. 3.3A. An important part of the 'circumstantial evidence' is that for the way chemical equilibrium behaves, the plots in A and B should be straight lines. The fact that they are is more important than the numbers then extracted from them.

Proceeding clockwise around Fig. 3.3, graphs C and D are for some additional experiments with another substance added to the culture medium, a good calcium binder called EGTA (ethyleneglycol tetracetic acid). Again, straight lines for calcium binding in C, with the line moved upwards by added EGTA. This means that EGTA is opposing (inhibiting) the effect of bound calcium on the spacing λ. In Fig. 3.2B, EGTA is represented as I (inhibitor) and shown as attaching to the R–C complex to make it inactive for patterning processes on the cell surface. Once again, one can extract a value for the binding constant of I to R–C. I don't give its value here because it proves little; a rather large range of values is possible for binding of a fairly large molecule to an even larger one. The point, again, is the linearity of the plots, the fact that the line in Fig. 3.3C is raised up in proportion to [EGTA] and that the expected linearity is found for the temperature dependence of the EGTA binding as plotted in D. These things are expected for what is known as an 'uncompetitive inhibitor', meaning that it attaches to a different place on R from the calcium C, but will not attach unless C is there. These types of plot can be found in any elementary textbook account of enzyme kinetics.

This completes my account of 'circumstantial evidence' for a dynamic mechanism for whorl pattern formation in *Acetabularia*. The fairly large space that I have devoted to experimental detail of one study represents my views that progress by such rather indirect methods remains important in science, and that it is tending to get overwhelmed by the kinds of work in which precise information on minute details can be found, usually with the aid of high technology: molecular structure determinations by X-ray crystallography and by NMR; sequencing of proteins, RNA and DNA, leading to complete listing of genomes and all the other–omics. It is easy to overlook the history of science of

200 years ago and forget that the atomic theory arose by indirect inference from the constant composition by weight of at least some compounds. (It wasn't even true of some others, leading to the dispute between Dalton and Berthollet, in which it was good for the quick advance of the atomic theory that Dalton's view prevailed, although Berthollet was sufficiently correct that we can now distinguish compounds into daltonides and berthollides.) When a general principle seems somewhere to be lacking, science often has to proceed by methods that seem to belie the term 'exact science'. It is applicable chiefly when the general principles are firmly established, and science can proceed to applications. That is the case today for molecular genetics. But to my mind, how the genetic information gets converted into macroscopic shapes remains a great unknown that is going to need a lot of 'circumstantial evidence' approaches, of which I have given one example. And it leads to a prediction for the biochemists: find that protein R in the *Acetabularia* plasmalemma.

3.2 *LARIX X LEPTOEUROPAEA* COTYLEDONS

Somatic embryos of this hybrid larch had been shown by Patrick von Aderkas (2002) to have a variable number of cotyledons within a clonal population; and addition of 6-benzylaminopurine (benzyladenine, BAP or BA) has been shown to bring about a decrease in their average number. In starting work on this whorl-forming phenomenon in collaboration with von Aderkas, I hoped to find a constant-spacing effect similar to that in the *Acetabularia* whorls, and to find that BAP was doing something to the spacing λ between cotyledon primordia. In the event, the first part of this expectation was fulfilled (Figs. 2.5D and 2.5E) and the second was not. It turned out that λ, as calculated from the slopes of Fig. 2.5E and the corresponding plot for culture medium with added BAP had, within experimental error, the same slope, from which λ could be calculated as 98 and 93 μm (both ± 4). So what we have is the constant-spacing phenomenon, and what more is there to be said about it that has not already been said about *Acetabularia*?

A feature that has interested me a lot, and that still needs both computational and experimental work, is the matter of possible mechanisms for formation of whorls of small numbers of parts. At the limit of 1–2 parts (monocotyledon embryos and dicotyledon embryos together with the multitudinous examples of dichotomous branching in plant development), it is clear that a single waveforming mechanism, whether reaction–diffusion or mechanical buckling or anything else, is capable of generating the pattern. For large numbers of parts, such as the *Acetabularia* whorls, I have maintained that a single mechanism is unlikely to be able to organize the primordial branches into a whorl. Indeed, the Dasycladales of half a billion years ago did not, but produced single hairs in random positions (Dumais and Harrison 2000). Over the next hundred million years, they learned the trick of organizing these appendages into a whorl. Computations made in my group with single reaction–diffusion mechanisms, and a wavelength short enough to put a large number of primordia on one growing tip have usually distributed them randomly. Hence, I have long advocated a sequence of two pattern-forming mechanisms to make whorled structures, the first to pattern a substance A in an annulus, the second to use that supply of A to make a whorl. This is the meaning of the stages labelled I and II in Figs. 2.1 and 3.1, and specifically indicated as reaction–diffusion in Fig. 2.1.

If such a two-stage mechanism is needed to organize large numbers, where is the lower limit? What kind of mechanism is the minimum, the thing to set up if Ockham's razor is hanging over one's head, for whorls of, say, four, five or six organs? Conifer embryos are among the few examples of variability in number when the numbers are small. For the larch hybrid somatic embryos, the plot of diameter d_2 versus number of cotyledons n_c, Fig. 2.5E, has a rather large intercept (c) on the d_2 axis, indicating that the spacing λ is established round a circle at a substantial inset $c/2 = 82\,\mu m$ from the edge of the flattened growing tip. Are the spacing λ and the circle at inset $c/2$ to be attributed to two mechanisms or one? If a single waveforming mechanism is to do it, the appropriate waveforms to consider are the patterns of vibration of a circular drumskin attached to a rigid outer border. These are a combination of circular up-and-down ripples with alternating up-and-down pie-slice patterns. The ripples are represented by mathematical functions known as Bessel functions (or cylinder functions, to translate from the German term). The pie-slice patterns are just simple sine wave patterns in the angular coordinate around the circle. The positions at which primordia of developing organs start to emerge will be defined by the places where Bessel function and sine wave both have their maxima. This way of approaching the question of how a pattern fits the disc is quite independent of what kind of waveforming mechanism is operative. It could be a reaction–diffusion mechanism, with 'up' and 'down' being levels of concentration of a chemical substance. It could be mechanical buckling, in which case 'up' and 'down' can be taken quite literally.

The following discussion of the fit applies to anything that will generate a 2-D waveform on a flat disc.

Fig. 3.5 shows: A: set of Bessel function ripples plotted against the radial coordinate of the disc (x). The function is called J_2. The subscript is the value of an index n, which corresponds to the number of cotyledons n_c in the present case. Each Bessel function will combine with an angular function with the same number n of maxima around the circle. In Fig. 3.5A, the zeros of the Bessel function are marked with values of an index $l = 1, 2, 3$, etc. Which of these is going to be at the outer edge of the disc? To arrive at a definite answer to this question would require commitment to a specific kind of wave-generating mechanism and then an extensive theoretical project on that one. In my group, we have had enough trouble with reaction–diffusion patterns on a hemisphere (Chapter 4) to avoid making rash statements about a disc. There is, however, one obvious simple possibility that is worth following up to see what kind of fit it gives. That is, that the edge of the disc is always at $l = 2$, putting just one up-and-down ripple between the centre and edge of the disc. The remainder of Fig. 3.5 shows this possibility. As the number n_c increases, the ripple is pushed more and more towards the outer part of the disc, with an increasing flat inner part. Compare Fig. 3.5B, C, D and E with the pictures of somatic embryos with various values of n_c in Fig. 2.4. More precisely, the distance from the maximum of the Bessel function to the outer edge of the disc is almost constant as n_c changes (distance between the two dotted lines, marked $c/2$). F shows a plot of radius versus n_c, oriented to show how it corresponds to the pictures above. Rotated anticlockwise through a right angle, it can be compared with Fig. 2.5E. A quantitative comparison can be arrived at as the value of slope/intercept for the plots. For the proposed waveform fit of Fig. 3.5, slope/intercept $= 0.226$. The solid line in Fig. 2.5E (experimental) has slope/intercept $= 0.19$; the broken line is drawn with the theoretical value 0.226. (Experimental values for the culture medium containing BAP gave a plot very similar to Fig. 2.5E with slope/intercept $= 0.21$.)

What is the significance of this good correspondence between one possible theoretical fit and the experimental data? Chiefly, it strengthens my usual contention that a pattern should be regarded as a whole, as one thing, one entity. Here, the thing to be explained by some mechanism is a drumskin-type waveform. The fit by itself does not point specifically to reaction–diffusion or mechanical buckling or anything else that could generate a waveform. But it does restrict the search to mechanisms that can do that.

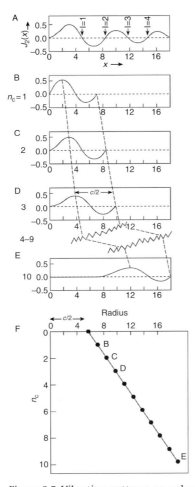

Figure 3.5 Vibration patterns on a drumskin are sets of circular ripples combined with angular functions (sin or cos, n_c cycles around the circle) to give overall patterns like those of n_c cotyledon primordia (Fig. 2.5D). Fig. 3.5 shows some of the mathematics of the circular ripples (Bessel functions, J_x, where x is radius from centre of the disc). A: the first three ripples, for $n_c = 2$. B–E: for $n_c = 1, 2, 3$ and 10, functions truncated after the first full ripple. If the outer edge of the disc does this truncation, the space between the two steep broken lines is $c/2$ as marked in Fig. 2.5D, i.e. the inset of cotyledons from the edge of the disc. F: the radius of the truncation after first ripple versus n_c. The broken line in Fig. 2.5E has the same ratio of slope to intercept as this plot, showing that the experimental data for hybrid larch cotyledons quantitatively match this set of Bessel function ripples. Rotate (F) 90° anticlockwise to compare it with Fig. 2.5E. From Harrison and von Aderkas (2004).

On the way to publication of this work (Harrison and von Aderkas 2004), we received a referee's comment that increase of cotyledon number with embryo diameter was 'unsurprising'. So: does a dog arising from a somewhat large embryo have six legs? Likewise, embryo length in a population of eggs of the fruit fly *Drosophila melanogaster* can vary by, commonly, about 30%; but all the wild-type embryos have three thoracic segments. In my previous book (Harrison 1993, pp. 56–7 and 334–7) I raised the question of measuring versus counting in morphogenesis as it happens and in theories of it such as reaction–diffusion. It is a question that one has to keep in mind continually when trying to put experiment and theory together. Nothing is unsurprising.

4
Dichotomous branching

4.1 BREAKING SYMMETRY

I return to the topic previously mentioned at the beginning of Chapter 2. To my mind, the essence of biological development is symmetry-breaking. I have expressed this view earlier (Harrison 1981, 1993, Chap. 5) and in the latter reference pointed out an apparent conflict between this and a view expressed by Meinhardt (1982): 'In most biological cases, pattern formation does not involve symmetry-breaking (although the proposed mechanism can perform this) . . . The asymmetric organism forms an asymmetric egg and the orientation of the developing organism is therefore predictable.' I countered with:

> Some of these [polarities] may be traceable back to the beginning of development (e.g. to the animal–vegetal polarity of an oocyte). This cannot, however, be the case for every polarity which is seen during development, for if all of these could be found at the outset, in proper spatial relationship, then, first, all development would be attributable to the inheritance of microscopic templates, and there would be no role left for the genome, and, second, the initial state would contain a miniature of the developed form. This would be equivalent to the long-discarded concept of the homunculus in the sperm.

There is no real conflict between these two views. In considering any developmental phenomenon that can be regarded as formation of a new pattern, one simply has to be careful about one's picture of the initial state and what features of it are changed or retained in the process under consideration. The changes may involve only amplifications of existing asymmetries, e.g. the modification of a shallow gradient to produce a big localized peak at the top end of it, which Gierer and Meinhardt (1972) called 'firing a gradient'. To this, and its probable relevance to positioning of localized single organs in animals, they gave much of their attention in their earlier work. But the changes can also involve both breaking of existing symmetries and making of new ones. These are dominant features of the mechanisms for generating patterns of repeated parts, a topic I have pursued particularly in plants.

For most of the present discussion of plant development, I start with an initial state, which is a hemispherical shell topping a cylindrical shell. The hemisphere represents the outermost layer of cells (tunica) of an apical meristem or, in unicellular cases, the active region of extension of the cell surface in tip growth. When I discuss patterns, e.g. of chemical concentration distributions on the hemispherical surface, I use spherical surface harmonics as the waveforms corresponding to simple sine or cosine waves for 1-D pattern, or to the vibrating drumskin patterns described in Chapter 3 for 2-D pattern on a circular disc. The simplest non-uniform pattern of this kind is in fact also just a cosine (of the co-latitude, if we take the extremity of the growing tip as pole and its junction with the cylinder as equator). These idealizations of the starting shape and of patterns upon it raise a number of other general issues to do with the kind of approach I am presenting in this book.

First, I am representing a growing region of a plant by a surface, a thin hollow shell. For multicellular plant development (and indeed for much of animal embryonic development) the active region is rarely fully 3-D. It is usually made up of a single layer of cells or at most two or three adjacent layers. For a plant meristem, one may consider chemical-kinetic patterning in the tunica layer only, or mechanical patterning in a tunica layer joined by 'springs' to the immediately subjacent layer. For single cells, the assumption that the cell surface is the important region in which pattern forms is much more questionable. Many biologists would automatically assume a priori that the shape of the cell is built upon a 3-D internal cytoskeleton, the forming of which must be addressed to explain morphological pattern. One of the examples to which a lot of attention has been given in the work of my group is the desmid alga *Micrasterias*, and I adduce evidence that the cell surface is the significant region in that case.[1]

Second, Green and King (1966) showed that the hemisphere is the steady-state shape for a growing tip if the growth rate of elements of the surface area goes with the cosine of the co-latitude. Paul Green told me that he had received from other biologists the objection: 'But the tip shape isn't a hemisphere.' Well, it isn't; see, for example, the drawings of *Acetabularia* tip shapes before the tips start to branch in Fig. 2.1B

---

[1] And see discussion regarding the *Acetabularia* surface: Section 3.1.2, esp. Fig. 3.4.

and 2.1C. But mathematical explorations of natural phenomena quite normally have to start from idealizations that permit the general nature of the explanation to be laid out in a relatively simple manner. Fine-tuning to get an exact match to a specific example can be added later. Even the concept of 'triangle' or 'square' or 'circle' is an idealization. If one were made, nobody would be able to see it, because it is made of lines of zero thickness. No living organism or other natural object can have circular or spherical symmetry, because these are defined by infinite-fold axes of rotation, and natural matter is finitely divided. (In Chapter 5 of Harrison (1993) I gave a more extended commentary on this kind of idealization under the heading 'Symmetry is in the spline of the beholder', together with other aspects of symmetry including the concept 'asymmetry begets asymmetry'.)

Third, pursuit of the mechanisms of patterning by computation rather than mathematical analysis also has to use a finitely divided representation of the shape of the surface, usually referred to as a mesh; and since plants are continually growing, the mesh has to have pieces added to it at frequent intervals as the computation proceeds. How to do this is very far from being a trivial problem. At intervals over a total period of about 25 years, I have had several postdoctoral fellows or research associates struggling with the problem of simultaneous patterning and growth of a surface region in 3-D space: Geoff Zeiss (Harrison *et al.* 1981), Mirek Kolář (Harrison and Kolář 1988), Stephan Wehner (Harrison *et al.* 2001) and David Holloway, who followed up 2-D work on repeated dichotomous branching (Holloway and Harrison 1999a) with the devising of computer software for 3-D plant growth that is available for anyone to use and is described in Chapter 5.

Fourth, while my preoccupation has been mainly with the possibilities of chemical-kinetic (reaction–diffusion) patterning; that of Paul Green was with the possibilities of mechanical buckling. He liked to start the discussion of symmetry-breaking by this mechanism from a flat disk.

4.2 MECHANICAL BUCKLING: THE POTATO CHIP ET AL.

If a flat circular disk has the same fractional rate of area expansion at every point upon it, then it will grow with an increase in size but no change in shape, if all parts of it continually adjust to relieve all mechanical stresses to zero. But something different must happen if the fractional rate is not the same over the whole surface. For example, if the rate increases with distance from the centre, the circumference will become too great in relation to the radius. Some deviation from the

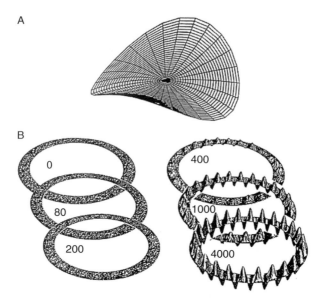

Figure 4.1 Paul Green's concept of whorl pattern formation by mechanical buckling at a plant apical meristem. A: computer-generated image by Dumais, for an informal 1998 symposium in honour of Green, of the 'potato chip', Green's favourite biomechanical analogue. B: computation of development of a 25-wavelength pattern by buckling of a ring. (See text; numbers from 0 to 4000 are relative times). With permission, from Green *et al.* (1996).

circular shape must then arise. The simplest is that the disk distorts out of plane into a 3-D shape with bilateral symmetry.

Fig. 4.1A is the mathematically idealized potato chip; it was the late Paul Green's favourite biomechanical analogue for the formation of a dichotomous branching pattern. But since it has not been generated by a distribution of growth rates with a circumferential maximum, it also illustrates Green's attachment to mechanisms involving mechanical forces directly. Beyond this simplest example, in collaboration with engineers (Green *et al.* 1996), he developed a theory of the formation of any number of spatially regular repeated structures by mechanical buckling; Fig. 4.1B shows the result of a computation of the development of a pattern of 25 wavelengths around a ring, with inner and outer radii 0.8 and 1.0 arbitrary units, and both inner and outer edges clamped. This boundary condition constrains the system to be approximately 1-D, and hence to distort at a single unequivocal wavelength. Similar patterns can sometimes be found on a vibrating metal plate (Waller 1961, especially patterns on cymbals, p. 11, and on a

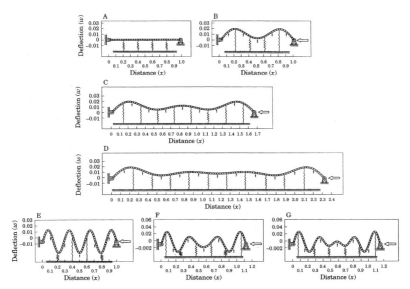

Figure 4.2 Computations of buckling of a straight bar with an elastic foundation (springs): *de novo* undulation with an intrinsic wavelength. The bar is compressed (arrow on the right). Both ends are held at zero displacement but hinged, a boundary condition analogous to a constant-concentration (Dirichlet) boundary condition for a chemical reaction–diffusion wave. B, C and D show buckling at the same wavelength with different numbers of waves for bars of three different lengths. E, F and G show bars of increasing length developing patterns of 3.5, 3.5 and 4.5 waves of slightly varying wavelength because the boundary condition requires the pattern to be a whole number of half waves. Between F and G, a 2% increase in length made the change from 3.5 to 4.5 waves. This is analogous to the behaviour of a chemical Turing wave as discussed in Section 2.3. With permission from Green *et al.* (1996).

vibrating plate, p. 109); but such a 2-D system can display any of a multitude of patterns of vibration from which this type would have to be selected when it happened to appear. The plant apex must produce it with the assistance only of its intrinsic 'Organizer'.

The model used to generate the patterns shown in Fig. 4.1 is illustrated in Fig. 4.2. A beam, of flexural rigidity D, can bend upwards or downwards, or both in different parts to form a wave-like shape, in response to the application of a compressive force in the horizontal direction. The beam is held down by a bed of springs, with force constant (or coefficient of elasticity) k (per unit area of the bed) to a rigid ground beneath. The correspondence to the plant apex is that the outermost layer of cells, the tunica, is the beam, and is regarded as elastically attached to the inner tissue, the corpus. The dimensions of D are [force][distance], and those of k are [force][distance]$^{-3}$, whence the wavelength expression, given earlier as Eq. (2.1c), $\lambda = 2\pi \, (D \, / \, k)^{1/4}$, correctly represents a distance.

Green's concept was that multicellular plants produce patterns of lateral organs at a shoot apex by this kind of mechanical mechanism, which he liked to style as 'biophysical' in contrast to 'chemical' mechanisms such as reaction–diffusion, which he tended to think of as modes of pattern formation more applicable to the surfaces of single cells. There, the surface may have mechanical properties as a viscoplastic layer, for which Dumais *et al.* (2006) have started to formulate a theory. This has a careful description of the properties of the cell surface in respect of deformability, but is presently set up for tip-growing systems in which circular symmetry about the axis of growth is not broken. The mechanical description of the surface lacks the spring-like tunica–corpus connection of the multicellular apices, and therefore does not have what it takes to be a prime mover in pattern formation; probably, chemistry will have to come in for that. I tend to look forward to an ultimate fusion of chemical and mechanical ideas in giving accounts of pattern formation at apices both of multicellular plants and of tip-growing single cells.

What may be involved in approaching such a fusion? The concept of a bending beam interacting with a spring mattress may seem far removed from that of chemicals becoming concentrated by reactions and spread out by diffusion. If we look further to the beginnings of existing mechanochemical theory, e.g. Oster *et al.* (1983) discussing patterning of mobile mesenchymal cells that move and interact by mechanical forces, we find a waveforming theory that, at least in the earliest stages of pattern formation, reduces to reaction–diffusion with the cell-as-molecule concept for the mobile cells (Harrison 1993, p. 136). For a cell diffusivity D_2, cell density N and mitotic rate r, the wavelength formula is:

$$\lambda = 2\pi(D_2/rN)^{1/4} \tag{4.1}$$

This is very similar to wavelength expressions for reaction–diffusion, with a rate of production of cells by mitosis replacing a rate of autocatalytic formation of molecules in a chemical reaction. But where is the unity between all these expressions and Eq. (2.1c), looking similar to reaction–diffusion expressions because someone has chosen to use the letters D and k, but in which these letters relate to bending of a beam and how many springs per unit area there are in a mattress?

Without need for further elaboration of existing theories, the unity exists and is quite straightforward. Every putative mechanism generating a quantitative scale of distance, pattern spacing or wavelength, contains one feature tending to increase the spacing and one tending to decrease it. The wavelength-increaser has the property of lateral communication across the pattern-forming domain; the wavelength-decreaser is entirely local, lacking all such communication. In reaction–diffusion, the lateral communication is by diffusion, while chemical reaction acts locally on the single-molecule scale. In mechanical buckling, the bending of the beam is laterally coordinated, while each spring in the mattress lengthens or shortens perpendicularly to the lateral direction and independently of all the other springs. In mechanochemistry of mobile cells, the movements of the cells depend on lateral mechanical forces between them leading to diffusion-like movements, while the mitotic division of each cell is individual and local. In all these mechanisms, a pattern spacing is achieved by a balance between a lengthener and a shortener.

Green *et al.* (1996), in their presentation of the mechanical buckling theory, compared this concept with reaction–diffusion, and wrote:

> The nature of cause and effect for both schemes, while familiar to physical scientists, is not commonly recognized in biology. In biology most causal chains involve many simple transductions whereby a compound is converted from one state to another (e.g. phosphorylated). The complexity is in the network. Here the complexity is within a single transduction. In the mechanism, balances of fluxes or forces are involved that require some calculus for their characterization.
> [Compare also my quotation from Wortis in Section 1.4.]

But besides appreciating an underlying unity at a certain level between these various concepts, we also have to seek by some experimental methods to determine which specific mechanism is operating in any given pattern-forming phenomenon. In the end, if that end is the completion of a bridge between the macroscopic theories and molecular biology, chemistry inevitably has to come in, whether in reactions involved in pattern-forming or in building materials with particular mechanical properties. At present, it is rather easier to approach the chemistry for the reaction–diffusion concept. For instance, in Section 3.1.2, I showed how an effect of calcium concentration on a pattern spacing could be interpreted physicochemically as a kinetic phenomenon, and that the quantitative character of this pointed towards a cell-surface calcium receptor. I do not know of any equally direct way to link any specific substance to its effect on mechanical properties that control spacing in a buckling mechanism. But this is not in any way an argument against the mechanism, just an indication that it may take longer to build the bridge. Meanwhile, Dumais has been looking rather directly for evidence on mechanical forces in shoot apices (Dumais and Steele 2000; Dumais *et al.* 2004).[2]

4.3 AN INTERLUDE ON FUNGAL HYPHAE: *NEUROSPORA CRASSA*

This section is an interlude in at least three senses: it is mainly about experimental evidence in a chapter that is otherwise devoted to theory; it concerns an example in which branching is not dichotomous in the wild-type organism, but becomes dichotomous in a mutant; and it involves a very low spatial size for compatibility with my favourite kind of mechanism. As mentioned in Section 2.2, the hyphae of fungi are long cylindrical outgrowths of cells, very fast-growing and commonly only about 1 μm in diameter, of the order of 100 times smaller than the diameter of an *Acetabularia* cell. And the discussion in Section 2.3 shows that, from the viewpoint of reaction–diffusion mechanism, this size for a pattern element is minimal, requiring for its control a very low diffusivity of order 10^{-12} cm^2 s^{-1}.

Hyphal growth is, however, perhaps the most extensively and intensively studied example of tip growth on single cells, from the viewpoint of trying to find its mechanism. Quite a lot is known about probably relevant chemical structures in the growing tip; but the cause and effect sequences for tip growth and for diverse variants of branching processes at and below the tip remain unknown. In contrast to both *Acetabularia* as discussed in Chapter 3 and *Micrasterias* as will be discussed in Chapter 5, there are known internal structures inside the apical region of the cell that are possibly relevant to the formation of branches. A growing tip is continuously supplied with vesicles, the surfaces of which are bilayer membranes that supply additional area to the plasma membrane upon fusion of the vesicle with it. Inside the polar region of the tip there is a small body called the Spitzenkörper (apical body). It consists of a cloud of vesicles, within which there is a smaller core region containing also other components such as microvesicles, ribosomes and actin microfilaments. It is always present when

---

[2] For more on mechanics in plants, see also Bohn *et al.* (2002); Goriely and Tabor (2003); Shipman and Newell (2005); Hamant *et al.* (2008).

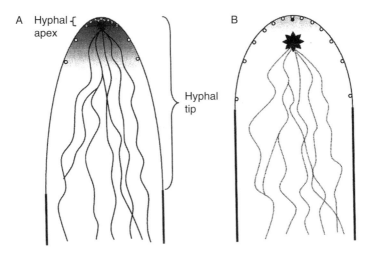

Figure 4.3 Schematic of structures within a hyphal tip of *Neurospora crassa*.
A: wild–type. B: *act¹* mutant. Circles = vesicles; wavy lines = actin
microfilaments; stars = accumulations of actin, etc. (Spitzenkörper); thin
outer line = expandable cell wall; thick outer line = non-expandable cell
wall; shading = $[Ca^{2+}]$. From Virag and Griffiths (2004) with permission.

a tip is growing, and may act as an organizing centre for supply of
vesicles to the membrane of the tip. It is always positioned below the
pole of the growing tip, and could be involved in the directionality of
growth. There is also a gradient in intracellular calcium ion concentration inside the apical region, with $[Ca^{2+}]$ highest just below the pole.
Fig. 4.3 summarizes the structure in a wild-type tip and a tip of an *act¹*
mutant.

Hyphae branch in both wild-type and mutant organisms; but the
wild-type branching consists of single lateral branches first appearing
a short distance (two or three tip diameters) below the equator of the
growing tip, while *act¹* mutant hyphae branch dichotomously in the apical
region (Fig. 4.4). When branching occurs, the original Spitzenkörper
disappears, and new ones form in the apices of the branches. But there
is no clear evidence indicating whether changes in the Spitzenkörpers
are causative for branching or an effect of it. The same can be said of
all the other components of the apical region; but the calcium gradient
seems very significant. Action of high $[Ca^{2+}]$ as an inhibitor of branching action could dictate the change in location of that action
between wild-type and mutant. Virag and Griffiths (2004) proposed 'a
calcium–actin–phosphoinositide (CAP) model in which actin regulates
the rate of vesicle flow from proximal to distal regions of a hypha.

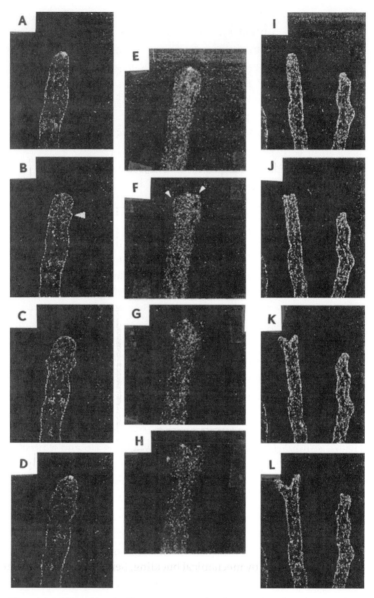

Figure 4.4 Branching in *Neurospora crassa* hyphae. Each column shows four successive stages for the same hypha, at equal time intervals. Left column (ABCD): time interval 75 s. Two lateral branches forming in a wild-type hypha. Middle column (EFGH): time interval 15 s. Aberrant apical dichotomous branching induced in a wild-type hypha by a high concentration (32 μM) of the vital stain FM4-64. Right column (IJKL): time interval 75 s. Apical dichotomous branching in an *act¹* mutant hypha. With permission from Virag and Griffiths (2004).

By doing this, actin controls the tip-high gradient of cytoplasmic calcium.' Their concept on branching is that a steep calcium gradient maintains apical dominance, while a shallow gradient leads to a broader spread of growth over the tip region and hence to the possibility of dichotomous branching. They mention also reaction–diffusion theory as a possible alternative, writing that an excellent candidate for a morphogen is calcium. Compare all this with my account of my concept versus Goodwin's (both reaction–diffusion) for whorl formation in *Acetabularia*, as given in Section 3.1.2, especially in relation to Fig. 3.4. The *Neurospora* evidence is closer to building the bridge between macroscopic features and genetics, because the act^1 mutation is expressed phenotypically in the act of pattern formation. That is why I have given a fairly large amount of detail to this instance. See also Section 5.4 for some reaction–diffusion computations relating to the general matter of maintenance of apical dominance while branching is happening.

4.4 PATTERNING ON A HEMISPHERE

I have already mentioned in Chapter 2 that spherical surface harmonics, the modes of vibration of a hemispherical bell, are actually quite familiar to scientists of whatever stripe (or other pattern) they may be, because all have had to learn enough chemistry to have had the hydrogen-like atomic orbitals thrust upon them. The angular parts of s-, p- d- and f-type orbitals, commonly shown in the usual kinds of sketches of the 'shapes' of the orbitals, are in fact spherical surface harmonics. They do not belong only to the mysterious world of quantum theory. One of the best introductory accounts of classical theory of vibrations in one, two and three dimensions is hidden in Chapter 3 of Kauzmann's (1957) *Quantum Chemistry*. There, he deals with the 3-D phenomena on a sphere by the example 'tidal waves on a flooded planet' (i.e. a sphere entirely covered with an ocean of uniform depth). For the present topic, the spatial waveforms might be actual outward and inward displacements of parts of the hemispherical shell by mechanical buckling, being the same as instantaneous snapshots of a vibrating hemispherical bell, or they might be concentration distributions of a chemical substance, generated and maintained kinetically out of an originally uniform spread. (In this sense, one can use negative values just as one would for inward mechanical displacements, because one may discuss the patterns as displacements from the initial uniform concentration.) Fig. 4.5 shows a representation of a hemispherical shell topping a cylindrical shell, and an indication of the meaning of the coordinates r (radius), θ (co-latitude) and φ (longitude).

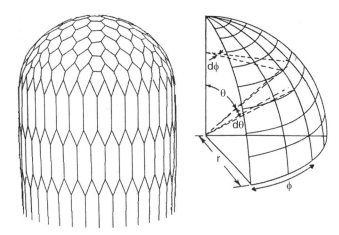

Figure 4.5 A representation of a hemisphere topping a cylinder by a simple mesh, and definition of the coordinates r (radius), θ (co-latitude) and φ (longitude) as used here for such a hemisphere.

At the equator, where the hemisphere meets the cylinder in the starting shape for discussing plant growth, I shall first assume zero displacement. This limits the range of spherical surface harmonics that one needs to consider. If the vertical axis is z, chemists and physicists who have in mind the atomic orbital analogy may forget about p_x, p_y, and all the d orbitals except d_{xz} and d_{yz}. Two indices, usually called l and m, are needed to specify any particular harmonic. A commonly used notation is $Y(l,m)$. The four lowest combinations of l and m values that satisfy the equatorial boundary condition are relevant to: continued tip growth with steady-state hemispherical tip shape (Fig. 4.6, column A); growth of a branch out of one side of a tip without a partner on the other side, as in the formation of the single seed-leaf of a monocotyledon (Fig. 4.6, column B); and all phenomena of two-fold symmetrical dichotomous branching, with an interesting aspect of why this happens rather than retention of circular symmetry by annular growth (Fig. 4.6, column C).

In each column, the first row is a sketch of the biological outcome of the developmental process. The second row shows a plan view looking down on the pole of a hemispherical tip, greyscaled for distribution of a chemical substance according to a spherical surface harmonic. (White is highest concentration, black lowest.) The third row is just to remind everyone where they have met these shapes before as the angular parts of atomic orbitals; in these plots in polar coordinates, the horizontal axis corresponds to the equator of the hemisphere, and only the upper half of the plot is relevant to the hemispherical tip. As already mentioned, the equator has been given a constant-concentration boundary condition (equivalent to 'Dirichlet condition', for the applied mathematician; 'nodal plane', for the quantum chemist), which, for $l = 1$, 2 and 3 in columns A, B and C respectively, limits the choice of patterns to those shown in Fig. 4.6.

It should be fairly evident that, if the patterned substance is a growth catalyst for the surface on which it is distributed, the patterns shown in the second row are going to develop into the morphologies shown in the first row. But what determines, in any particular biological system with a roughly hemispherical growing region, which of these patterns will arise?

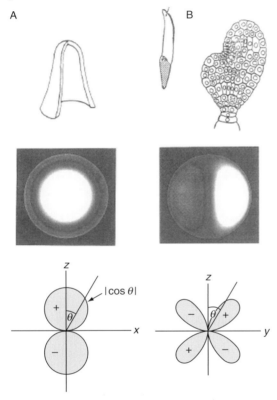

Figure 4.6 Plant growth phenomena and relevant spherical surface harmonics for a hemispherical surface as the starting shape (see text for full description).

Top row	Middle row	Bottom row
A: tip growth	A: harmonic Y(1,0)	A: p_z or p_o
B: monocotyledon embryo	B: harmonic Y(2,1)	B: d_{xz}
C: dichotomous branching, e.g. dicotyledon embryo	C: harmonics Y(3,2), Y(3,0) and sum	C: f orbitals

C

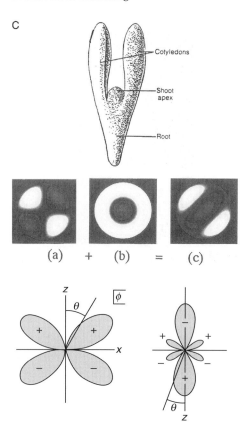

Figure 4.6 (cont.)

Let us go back to one dimension, as in Fig. 2.6. There, three alternative patterns labelled A, B and C differed only in wavelength λ. What is the analogue of λ for the alternatives A, B and C in Fig. 4.6? This question needs the mathematics of Turing dynamics, for instance as given in Chapter 7 of Harrison (1993). There, the dynamics leading to expressions for growth or decay rates of pattern (k_g), etc., were given in terms of a quantity $\omega^2 = 4\pi^2 / \lambda^2$, for 1-D pattern. (Physicists are very familiar with k^2 for this quantity. As a chemical kineticist who wants to use k symbols all over the place for chemical rate constants, I chose to use ω, which is more commonly used in connection with the time-dependent rather than the space-dependent part of vibrations.)

The corresponding mathematical treatment for a spherical surface of radius r shows that the correct expression for k_g is obtained simply by replacing the above meaning of ω by $\omega^2 = l(l + 1) / r^2$. This implies that for a pattern specified by any particular value of l there will be a 'band-pass' range of radius r just like the range of λ with positive k_g in Fig. 2.6. The pattern alternatives A, B and C in Fig. 4.6 will arise in that order for increasing tip radius. This becomes particularly interesting for pattern C. Here, unlike A and B, there are two quite different patterns for the same value $l = 3$: (a) a dichotomously branched pattern; (b) an annular pattern. So which is the plant going to produce? The mathematics has told us that all patterns of the same value of l should grow at the same rate k_g. The growth rate does not depend on the other index m of a spherical harmonic; we are here concerned with (a) $Y(3,2)$ and (b) $Y(3,0)$. Both of these patterns should grow together, giving an overall pattern (c), which is just the sum of (a) and (b) in equal proportions. This looks a good pattern to correspond to many instances of dichotomous branching, e.g. in the dicotyledon embryos. (The $l = 3$ harmonics that correspond to this simplest form of branching are, in the atomic orbital analogy, the angular parts of f orbitals, so rarely needed by the majority of chemists that the shapes of them are not shown in most chemistry textbooks.)

This discussion has shown that the selection, at a growing tip or apical meristem, of continued circularly symmetric tip growth, asymmetric outgrowth of a single organ, or dichotomous branching can be accomplished in reaction–diffusion dynamics according to the fit of parameter values (rate constants and diffusivities, which determine the optimum wavelength, or ω^2) to the tip radius r. This has been done using the mathematics of spherical waves, but without computation except for illustrative purposes in drawing pieces of Fig. 4.6, e.g. turning the underlying trigonometrical functions into greyscaled pictures. At a northwest regional meeting of the Society for Developmental Biology (USA), I talked on this topic, and raised the question: 'What could scientists be doing on this today if biology had developed as it has but the computer had not been invented?' and remarked 'I have just given you that lecture.' The remark seemed to generate above-average applause. (I am here referring to the computer as an instrument for scientific calculations, not as a replacement for the typewriter and other business machinery.)

4.5 SO WHY DO WE NEED TO DO COMPLICATED COMPUTATIONS?

The approach to pattern formation I am giving most attention to in this book is all to do with non-linear dynamics. When these arise in rates of chemical reactions, non-linear most commonly means that second- or third-order reactions, involving squares or cubes of concentrations, or two or three concentrations multiplied together, appear in the rate equations. Pattern-forming behaviour is intrinsically dependent on the presence of these non-linearities. But it is usually possible to find as a possible outcome of the dynamics a spatially uniform steady-state. In pattern-forming conditions, this will be unstable; non-uniformities will develop, at first in very low amplitude. At that stage, the only terms that matter in the rate equations are linear, containing only first powers of concentrations. Other, non-linear, terms are so small that they can be ignored. It is at this stage that the patterns can be represented by spatial waves: in 1-D, sin or cos, as in Fig. 2.6; in 2-D, sin or cos together with Bessel functions, as in Fig. 3.5; and in 3-D, the spherical surface harmonics, as in Fig. 4.6. Each of these will grow or decay exponentially with time, usually so that one of them comes to dominate the pattern; but there can be cases like that of Fig. 4.6C, in which two harmonics grow at the same rate and contribute equally to the final pattern.

But exponential growth cannot go on forever; it has to slow down. In terms of the equations representing the dynamics, it is the non-linear terms that bring about the slowing down. As pattern amplitude grows, they can no longer be ignored. And they lead to substantial distortions of the pattern from the simple waveforms. This is discussed in Chapter 6 and illustrated, for 1-D pattern, in Fig. 6.9. There, a distinction is drawn between 'primary patterning', which is the formation of harmonic wave-forms, and 'secondary patterning', which distorts those shapes. Precisely what this latter part of the process looks like is usually not accessible to mathematical analysis.

One can arrive at a good idea of the patterns that a reaction–diffusion mechanism will generate by linear analysis without compu-tation: first, if there is good reason to believe that pattern is morphologically expressed while it is, chemically, in the primary stage; second, if one is more concerned about wavelength than about the detailed shape of the final wave. The indication in Fig. 6.9 that a wave can distort in shape but keep its original wavelength is a good indica-tion of the behaviour of some (but not all) non-linear dynamics, espe-cially in 1-D.

But there are many instances in which computation is essential. Particularly, if two or more harmonics compete on an equal basis in the primary patterning stage, what is going to happen in the secondary stage? We have already seen that on a hemisphere the two waveforms that grow equally in the primary patterning when r is appropriate for $l = 3$ cannot be described simply in terms of a wavelength: $Y(3,0)$ is annular, $Y(3,2)$ branched. If patterning goes into the secondary stage, how does each of these distort, and what happens to their sum? The distortion, by the way, is represented mathematically by adding in other terms than those two; analysis of some of the computational results of my postdoctoral fellow Stephan Wehner showed a significant contribution from $Y(5,4)$. Not only does such patterning have to be studied by computation, but for this kind of competitive situation between two patterns, one has to be suspicious about whether the outcome is always going to be the same. This requires a statistically significant number of repeat computations, with some attention to not putting exactly the same random noise into the start of each one. For groups with moderate resources (and trying therewith to do both experiment and theory), this level of statistical luxury became achievable only in the course of the 1990s. Wehner, from his computer in Vancouver, was often running repeats of this kind on machines in two other buildings across campus and another somewhere in Montreal to present me with the results of 100 runs. And he did this for three different reaction–diffusion mechanisms, with contrasting non-linearities: the Prigogine Brusselator, the Gierer–Meinhardt model and my own hyperchirality model (see Chapter 6 for a description of these). The result was, roughly, that for each of these models run at optimal conditions to generate $l = 3$ patterns, the final pattern was dichotomously branched in about 80% of the computations and annular in the other 20% (Harrison *et al.* 2001).

     Such variability in developmental behaviour is quite common in a set of growing tips or apical meristems. For example, in the work of von Aderkas (2002) on a clonal population of larch somatic embryos, which led to the study described in Section 3.2, about 20% of the embryos failed to form cotyledons, usually halting at a stage of annular patterning. Is such variability an intrinsic property of the type of dynamics explored in our work, reaction–diffusion? This question led to a collaboration with an applied mathematician, Wayne Nagata, who showed by bifurcation analysis of the dynamic equations that variability quite close quantitatively to this 80/20 split was indeed to be expected for the two types of non-linear dynamics he studied (Brusselator and hyperchirality) (Nagata *et al.* 2003). Therefore, in a population in which the majority behaviour is a branching event, the existence of a minority that retain circular symmetry at that stage does not necessarily imply the existence of a mutation (e.g. the 'cup-shaped cotyledon' mutants CUC1 and CUC2, Aida *et al.* 1997). This could lead me into a discussion of the limits of variation to be expected in development in a population in the absence of mutants and even in the absence of variability in genetic composition, i.e. in a clone. I return to this in Chapter 8 in relation to early fruit fly development. For the present account, we have to cope with the fact that plants keep on growing continuously not only in simple extension of a stalk or root but also when new organs are forming, and the hemispherical tip isn't going to stay hemispherical very long. The pursuit of any kind of theory set up to relate patterning of growth to development of form demands complex computations. This is exemplified by the work described in the next chapter.

Meanwhile, the foregoing discussion raises a rather general question about chemical patterning and its expression in morphological change. It is quite common to assume that the former is much faster than the latter, so that a reaction–diffusion computation should be run to steady-state to find the pattern that governs change of shape. This is not necessarily so in all cases. Where dichotomous branching is closer to 100% robust than the 80% instances mentioned, it is likely that the pattern is the equal mix of harmonics $Y(3,0)$ and $Y(3,2)$ described in Section 4.4. That is, a pattern maintained through a large part of the formation of a reaction–diffusion pattern, but distorted before steady-state is reached. The earlier stage is likely to be grabbed and expressed by what is seen morphologically, even if chemical pattern later distorts. In addition to plant development, instances include many of the superficial pigmentation patterns on sea shells discussed in Meinhardt's (1995) book. I return to this topic in Chapter 5 (in regard to Eq. (5.3)) and in Section 6.3.4.

5

Micrasterias, and computing patterning along with growth

In this chapter, especially in Section 5.2.2, I give a fair amount of detail on how one goes about a computational project to explore the ability of reaction–diffusion mechanisms to generate a sequence of dichotomous branchings as they are observed in a rather complex morphogenesis of a single cell. Several generalities should be borne in mind by the reader: first, we are not merely pursuing the objective of the animator and many practitioners of computer graphics to draw a good moving picture of something no matter how we do it. Our objective is that of the developmental biologist: to draw the plant and its development the way the plant does it.

Second, many readers, including experimental biologists, may have no clear concept of what it takes to do the theoretical side of a project in developmental biology. The message here is that, in proper pursuit of the scientific method, experiment and theory are about equally time-consuming. I hope that Section 5.2.2 illustrates how the theoretical work requires groups with graduate students and postdoctoral fellows fully devoted to it, just as biological experimentation does.

Third, as one pursues such a theoretical project, one gradually begins to realize more and more that negative-looking features of the postulated mechanisms are essential to getting the right development of shapes. For plants with many small and separate growing regions, how their boundaries are established and how they manage to move to keep the active regions small is often more important than what is going on inside the active regions. And inhibitory effects, or depletion of reactants in chemical processes, are often more important than formation of products. As we put it in Holloway and Harrison (1999a) in regard to movement of boundaries limiting the active regions: 'Our general conclusion is that this apparently negative aspect of morphogenesis is crucial in governing species-specific cell architecture.

77

The self-organizer of the cell morphology is perhaps more of a self-sculptor than a self-constructor.'

Fourth, we may seem to pursue reaction–diffusion as if it were the be-all and end-all, ignoring the fact that many developmental biologists doubt that it has any relevance to their discipline. But one has to start a theoretical exploration with some means to progress, and reaction–diffusion is still the only fully developed theory with adequate pattern-forming power. Some conclusions that we reach are quite likely to be much more general than the specific type of theory we have used to arrive at them. For instance, the concept that the production of repeated dichotomous branchings in the developing shape of a surface probably needs an intracellular timing mechanism to tell the surface when to make particular changes is probably much more general than reaction–diffusion.

5.1 PATTERNING AND ITS MORPHOLOGICAL EXPRESSION

Chapter 4 has shown by example that there has been and remains plenty of work for mathematically and computationally inclined scientists to do to establish the possible scope of non-linear dynamics in application to developmental phenomena, even when the exploration is confined to a region of fixed size and shape. Such complete separation of the patterning process from its morphological expression in change of shape can be quite relevant to many developmental phenomena, especially in the animal kingdom. Therein, an embryo has to undergo a host of developmental events before it has made for itself an efficient feeding system beyond simple contact of cells with yolk. Conservation of size is useful in early events. In some of the most extensive studies of reaction–diffusion, e.g. the books by Nicolis and Prigogine (1977) and Meinhardt (1982), it has been explicitly indicated that the work has been for patterning in regions of fixed size, shape and boundary conditions.

Plants, however, commonly grow continuously throughout their lives, with very big changes in size and shape, and frequently renewed organogenesis that the Organizer would not contemplate in an adult animal; as Aristotle noted, he's long gone (see quotation in Section 1.1). Although it remains difficult to get an unequivocal experimental measure of how long a patterning event takes for most patterning events in most kinds of organism, it is reasonable to suppose that patterning and its morphological expression are not neatly separated in plants, and that one should try to cope with the possibility of their continuous mutual interaction. This provides a more cogent reason for doing elaborate computations than I gave in the preceding chapter. Even if growth is going rapidly enough to pick up what I there called the primary stage of patterning, which is accessible to mathematical analysis, it is still beyond the capabilities of analysis to cope with patterning in a domain of size and shape continuously changing in very diverse ways. Computational treatment is essential.

In my group, we have returned to this problem from time to time over quite a long period, with increasing sophistication in how the pattern-morphology interaction was tackled. Zeiss (in Harrison *et al.* 1981) made our first primitive attempt to combine patterning and morphological growth within one computation by studying, computationally, patterning on a hemispherical shell with radius expanding uniformly in time (Fig. 5.1). Two reaction–diffusion models were used in hierarchical series: the first produced annular concentration maxima of a substance A, but included repeated generation of a central maximum (resumption of tip growth) as the hemisphere expanded. The second stage produced periodically repeated maxima of a substance X around the annular maximum of A, i.e. whorl pattern. Kolář (in Harrison and Kolář 1988) tackled growth not uniform in time but resulting from catalysis of area extension in proportion to the concentration of a patterned substance X. He computed in 2-D (growth of a planar closed loop) and succeeded in generating branching patterns of development, but always with a large (obtuse) angle between the two lobes formed in a dichotomous branching event. Holloway (in Holloway and Harrison 1999a), still computing in 2-D, added extra features to the mechanism that got control of the branching angle, allowing the formation of lobes with acute or obtuse angles between them according to parameter values; and he then extended the computations to 3-D. But my long-time collaborator Thurston Lacalli once remarked to me: 'It's the simplest computations that give the most information.' So, Adams and I have recently returned to 1-D, thereby simplifying the computation programs because they increase greatly in complexity with increased dimensionality of growth. But this lets us play with increased complexity in the chemical mechanisms explored.

5.2 *MICRASTERIAS:* DEVELOPMENTAL FACTS AND 2-D COMPUTATIONS

5.2.1 Developmental facts

The desmids are an abundant family of algae, found in fresh water worldwide. A number of genera display tip growth at some stage in their development. Repeated dichotomous branching, however, is

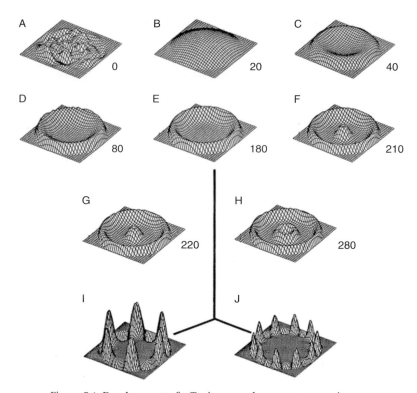

Figure 5.1 Development of a Turing morphogen concentration wave on a hemisphere, which is mapped onto a circular disc inscribed in a square. Distances from the centre are distances along a meridian. The plane of each diagram is homogeneous steady-state morphogen concentration. Vertical relief shown in perspective is concentration of the morphogen. Numbers are relative times. Sizes are normalized, i.e. the hemisphere is actually growing linearly in time. A: random input. B–H: tip growth pattern, to annular pattern, to the same plus renewed tip growth, as the hemisphere expands. At E a second stage (feedback loop II in Fig. 2.1C or 3.1) is arbitrarily switched on, giving whorl formation around the annulus. Diameters are in the ratio of 6:10 in I and J, showing constancy of spacing or proportionality of number of organs in a whorl to diameter. Computations by Zeiss. From Harrison *et al.* (1981) with permission.

most clearly manifest in the genus *Micrasterias* (e.g. Prescott *et al.* 1977; this volume, Fig. 2.2); and each branch becomes a growing tip. Much of the marked interspecific variation of form within the genus lies in difference of branch width, length, total number of branching events and branching angle. In the polar lobe (Fig. 2.2A) this angle is usually large, between 70° and 180°, depending on species. The wing lobes have much smaller branching angles: from approximately 30° to 60° in the first branching, and down to about 5° in some subsequent branches, depending on species. The diversity of cell shapes that these variations give rise to (Fig. 5.2; Couté and Tell 1981), with many of the details relied upon for species characterization, exemplifies very clearly that to decide upon a type of mechanism (e.g. 'Brusselator reaction–diffusion') as the probable prime mover in generating branching is only the beginning of producing a reasonably exact match between observed shape and shape computed from the mechanism. Much detail has to be added painstakingly to pursue the scientific method as described by Crombie (1959, vol. II, p. 288):

> In the actual history of science many of the most fruitful theories have been developed from preconceived ideas of the kinds of laws or theoretical entities that will be discovered to explain the phenomenon. The history of the enquiry has to a large extent consisted of using the sharp tools of mathematics and experiment to carve out of these preconceptions a theory exactly fitting the data.

Experimental data obtained on *Micrasterias* over several decades led, by the 1980s, to a fairly detailed picture of the cellular biology of morphogenesis (Kiermayer 1981). At mitosis of the adult cell, a septum divides it into two daughters which remain physically joined at the septum, though biologically developing as two separate cells (Fig. 2.2B). The septum of each blows out into a bubble of primary cell wall, thought to contain already some rudimentary template for patterned growth (Kiermayer 1964, 1967, 1970). In the course of 4–5 h, these bubbles grow and undergo the repeated dichotomous branchings that generate the adult form. (The most detailed information available is for *M. rotata*; the number of branchings differs in different species.) The new cell surface is delivered from the Golgi bodies as 'dark vesicles' that fuse with the plasma membrane, supplying primary wall precursors (largely pectic), membrane and cellulose synthetase rosettes (Giddings *et al.* 1980). Since each rosette gives rise to a microfibril of fixed length (Staehelin and Giddings 1982), it would seem that extension of wall must go in lock-step with extension of membrane.

In experiments with reduced turgor pressure, the cell surface no longer extended, but wall material continued to be delivered and accumulated on the inside of the plasma membrane (Kiermayer 1964; Tippit and Pickett-Heaps 1974; Ueda and Yoshioka 1976). Accumulation was heaviest at the growing tips, and clearly reflects the morphogenetic

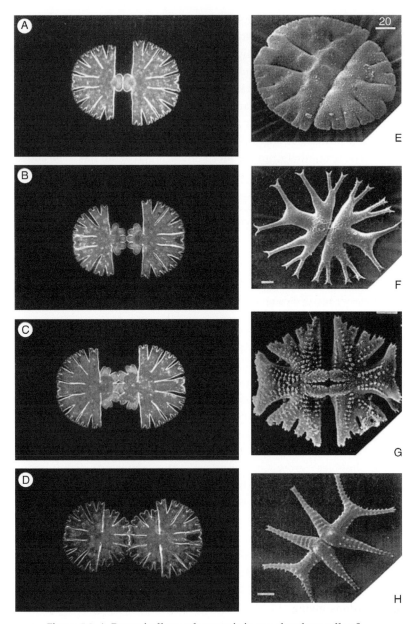

Figure 5.2 A–D: semicell morphogenesis in two daughter cells of
Micrasterias thomasiana, roughly 1, 2, 3 and 4 h stages, from Harrison (1994)
with permission. E–H: SEMs of fully formed cells illustrating diversity of
form within the genus: E – *M. verrucosa* var. *verrucosa;* F – *M. radiata* var.
brasiliensis; G – *M. americana* var. *bimamillata;* H – *M. tropica* var. *senegalensis.*
The bars indicate 20 μm in E, 10 μm in F–H. From Couté and Tell (1981)
with permission.

pattern. Fluorescent labelling of membrane-bound Ca^{2+} showed a corresponding pattern, with highest Ca^{2+} on the growing tips (Meindl 1982). This is in accord with the role of Ca^{2+} in fusion of vesicles to the plasma membrane (Gratzl 1980; Steer 1988; Battey and Blackbourn 1993). Ca^{2+} patterning is indeed limited to the plasma membrane, or at the most to a very thin submembrane layer; using microinjection of the fluorescence chelate fura-2, which detects free cytoplasmic Ca^{2+}, Holzinger *et al.* (1995) found no cytoplasmic gradients of Ca^{2+}.

Such experimental data point towards the plasma membrane as the site of control of patterned surface extension. Internal cytoplasmic patterning is further contra-indicated by the absence of any relation between cytoskeletal elements and the patterning: disruption of micro-tubule assembly by colchicine has long been known to have no detect-able effect on morphogenesis (Kiermayer 1968); and while disruption of actin microfilaments by cytochalasin B inhibits normal development (Ueda and Noguchi 1988), there is no indication that microfilament arrays initiate branching. There seems to be consensus in the experi-mental literature that pattern initiation happens at the plasma mem-brane. As far back as 1969, Thurston Lacalli chose morphogenesis in *Micrasterias* as the topic of his PhD thesis (Lacalli 1973a, 1973b, 1975), largely because it appeared to be an instance of control at the surface of its own rate of extension in area. His work attracted my attention to the entire field of morphogenetic mechanisms (preface to Harrison 1993) and led to our collaboration in a first attempt at explanation by kinetic theory and computer modelling (Harrison and Lacalli 1978).

5.2.2 2-D computations

Another attractive feature of *Micrasterias* for computer simula-tions of development was that it was possible to make a first shot at it (and, in the event, two more shots, over a period from the 1970s to 1999) using only two spatial dimensions. What makes this possible is that the cell surfaces become rather flat, with something approximating a sharp edge, so that the final shape resembles a biconvex lens with deep indentations giving the stellate form (Fig. 5.2). During development, each new lobe in fact advances by tip growth, so that the most active region is a dome which should, strictly, be treated three-dimensionally. But in practice, the 2-D growth of a longitudinal cross-section through the top of a dome is a good approximation of the advance of the entire dome with the same distribution of rate of surface extension, if the growth is isotropic. Slow-growing regions that become the bottoms of

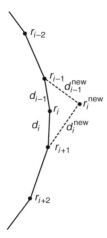

Figure 5.3 Geometrical basis of the 2-D algorithm (see text). From Harrison and Kolář (1988), with permission.

the indentations in the cell outline are quite a different matter, and required an arbitrary addition to our 2-D growth algorithm (Holloway and Harrison 1999a) to prevent them from continuing to move outward when there was no growth at these points. The proper treatment of such saddle points demands 3-D work as described in Section 5.3. But the combining of a chemical model for patterning with its morphological expression is much more efficiently explored in the first instance in 2-D. For *Micrasterias,* work in this approximation was sufficient to show us what variations in parameter values and features of the model were needed to reproduce in computations the diverse final shapes of various species in the genus.

In all the work done in my group on this problem, the assumption has been that area extension of the cell surface is catalysed in proportion to the concentration of a patterned substance, either X or Y in a Turing-type reaction–diffusion mechanism that has done the patterning. (It is X in all the work described in this section.) In 2-D computations, the surface is represented by a number of points (nodes, in the usual terminology of computations) with specified positions r_i, which are envisaged as joined by straight lines of lengths d_i (Fig. 5.3). The chemical patterning is translated into its morphological consequence by increasing each d_i according to the concentration of catalyst X, and using intersections of circles centred on nodes $(i-1)$ and $(i+1)$ to find the new position of node i.

The kind of chemical mechanism pursued in this modelling is reaction–diffusion in the general form capable of generating concentration distributions as stationary spatial waveforms, as discussed in the preceding chapters; the patterning on a hemisphere discussed in Section 4.4 is particularly relevant. If one takes the simple cosine pattern as in Fig. 4.6, column A, and takes a longitudinal cross-section of it, one obtains a semicircle with a 1-D cosine pattern on it. In our 2-D growth algorithm, this would advance upwards at constant shape to elongate two vertical straight lines below it exactly as the hemisphere would advance to elongate a cylinder below it (because of circular symmetry). There is no approximation in this idealized (axisymmetric) example; the correspondence between the 2-D and 3-D case is exact.

The patterned substance is envisaged as acquiring that distribution dynamically because it is one of two substances (commonly styled X and Y) in a two-morphogen mechanism of 'activator–inhibitor' or 'activator–depleted substance' type. These were the dynamics first proposed by Turing (1952), which have subsequently been elaborated in diverse forms: see Chapter 6 of this book and Chapters 7 and 9 of Harrison (1993) for some review of this diversity. Not every one of these forms is equally suitable for every example of morphogenetic patterning. For branching processes, particularly in relation to specific instances in plants (including algae), my group has generally leant towards the Brusselator (Prigogine 1967; Prigogine and Lefever 1968) or variants of it (Tyson and Light 1973; Tyson and Kauffman 1975) as having the most appropriate properties. As a matter of historical completeness, Harrison and Lacalli (1978) first tackled *Micrasterias* using a reaction–diffusion model of my own devising, called 'hyperchirality', which has interesting symmetry properties but is probably not the best for this example of morphogenesis. Lacalli (1981) continued the work on application of reaction–diffusion to unicellular algae using a Tyson modification of the Brusselator. In subsequent work in my group, the original Prigogine Brusselator has been used throughout, albeit with some additions to the mechanism.

The modifications made to the Brusselator for the *Micrasterias* project illustrate fairly precisely the main approach that I advocate in this book for a physicochemical approach to pattern-forming events in biological development. The original mechanism (Prigogine 1967; Prigogine and Lefever 1968) was most conveniently formulated as a chemical reaction mechanism in four steps, involving reactants A and B, intermediates X and Y which can become patterned in distribution, and final products D and E which are of no consequence for the patterning phenomenon:

$$A \rightarrow X \quad \text{rate constant } a \tag{5.1a}$$

$$B + X \rightarrow Y + D \quad \text{rate constant } b \tag{5.1b}$$

$$Y + 2X \rightarrow 3X \quad \text{rate constant } c \tag{5.1c}$$

$$X \rightarrow E \quad \text{rate constant } d \tag{5.1d}$$

X and Y must also be diffusible (diffusivities D_x and D_y, with $D_y > D_x$).

The roles of X and Y in steps (b) and (c) hold the essence of the pattern-forming property; but the reactants A and B have powers for substantial modifications to the pattern that will be formed, as also do the diffusivities. For example, if a sinusoidally periodic pattern forms, it has a wavelength λ, and this is quantitatively determined as

$$\lambda = 2\pi \left[D_x(n-1) \middle/ \left\{ (n+1)(A^2 B/n)^{1/2} - A^2 - B + 1 \right\} \right]^{1/2}. \tag{5.2}$$

Here, italic A and B are the concentrations of substances A and B, the rate constants of the four steps have all been assigned the value unity (a convenience used extensively, e.g. in Nicolis and Prigogine 1977), and $n = D_y/D_x$. (Cf. Eq. (5.2) with general form of Eq. (2.1a).) It is evident that, if the spacing between repeated parts in a developmental pattern is the wavelength λ of a Brusselator, it is under the control of the concentrations A and B of the inputs to the mechanism, which are not themselves the patterned substances.

The reactants A and B may, however, be involved in more complex aspects of pattern control. One of these is the aspect of tip growth described in Section 3.1 and indicated in Fig. 3.1 as needing a mechanism designated 'feedback loop III'. This is the 'skyhook' aspect of tip growth, in that the growing region must limit its own size by pulling its boundary up after it as it advances. This seems to mean that either the concentration of one of the inputs A and B, or the rate at which it is converted into X or Y must be under the control of the X, Y pattern. Various possibilities for this have been and are being studied in my group (Harrison and Kolář 1988; Holloway and Harrison 1999a). In the 2-D computations I describe in this section, an earlier stage was postulated in which reactant A is formed out of a precursor S with X as a catalyst, and A also undergoes second-order decay:

$$\partial A/\partial t = k_p SX - k_d A^2 - aA. \tag{5.3}$$

But to complete the boundary condition so that patterns of repeated concentration maxima and minima lead to isolation of the active regions from each other (the general characteristic of plant development), we had to add a condition that, wherever X fell below a specified threshold value (X_{th}), the entire patterning mechanism ceased to operate, except for the final decay of X expressed by Eq. (5.1d). At this stage, we did not seek to explain the existence of X_{th} by yet another putative chemical mechanism. (The general concept of this additional feedback loop has some resemblance to the 'extinguishing system' in Meinhardt's (1995, p. 111, Eq. 7.1) work on a particular cone shell pigmentation pattern.) To get rid of the arbitrariness of the threshold assumption is part of our continuing work, and continues to use features of the original Turing equations; they're very versatile.

The kinds of additions just listed to the Brusselator scheme typify the sort of work that is done by 'modellers' (i.e. 'theoreticians' in the earlier conventions of description of the scientific method) to devise mechanisms having the power to carve out precise morphologies displaying species-specific variations within a genus. The linkages present within these mechanisms may include both hierarchical relationships in which one developmental event significantly affects the outcome of the immediately following event, and feedbacks in which an event affects itself through influencing its antecedent events. For the latter, a particularly important instance in the present case is the control of branching angle by that same process that works to separate growing tips from each other. Harrison and Kolář (1988) devised a putative ageing effect of the cell surface, which was not a mechanical stiffening of the wall but a chemical decay of the rate of formation of X from A, i.e. aA, perhaps by decay of a membrane-bound enzyme controlling the rate constant a. (It should be recognized that Eq. (5.1a–d) imply not only the existence of substances A, B, C, D and E, but also most probably the existence of four enzymes governing the rate constants a, b, c and d.) Where the cell surface is growing rapidly (regions of high growth-catalyst concentration X), new enzyme is continually supplied, rejuvenating the surface; where there is no growth, the surface is ageing according to the conventional meaning of that word, in proportion to time. On this basis, Kolář was able to start from a circular loop and compute morphological change by successive dichotomous branchings, but always with the branches at obtuse angles (Fig. 5.4b). This is a good start, but inadequate to explain *Micrasterias* morphogenesis, which always displays some acute-angled branchings, e.g. Fig. 5.4A.

Holloway used instead the feedback into production of A described by Eq. (5.3), together with the X_{th} concept for irreversible cessation of the patterning mechanism at low X. This kind of feedback proved very powerful, needing only a few months of intensive work on

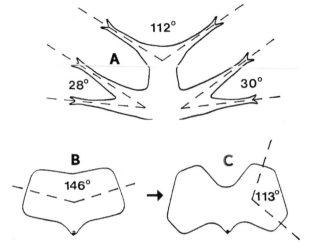

Figure 5.4 The problem of devising a theory to explain a variety of branching angles. A: obtuse and acute branching angles in M. *radiata* var. *radiata*, reproduced from Prescott *et al.* (1977), by permission of the University of Nebraska Press. © 1975 by the University of Nebraska Press. B and C: two stages from a computation of Harrison and Kolář (1988), with permission, producing only obtuse angles for both first and second dichotomous branches.

the computer to persuade it to generate acute branching angles. The general concept from which this aspect of the work started was that the concentration pattern of growth catalyst X, which leads to growth of two branches, at first leads to tip flattening (Figs. 5.5A and B). This will continue into high-angle branching (Fig. 5.5C) unless the feedback controlling the boundary of the tip advances that action so that the take-off regions for the start of the branches are pushed up onto the flattened area (Figs. 5.5D and E), simultaneous with killing growth in the newly formed trough between the branches.

With these sorts of additions to the Brusselator mechanism, we were able to perform computations matching fairly well a number of the diverse shapes displayed by various species of *Micrasterias* and one of the related genus *Euastrum* (Fig. 5.6). Despite all the additional features, however, our theory is still incomplete. To generate different shapes, the operator had to intervene at particular times in a computation and change the value of the threshold X_{th}. This parameter value (controlling feedback loop III) is very powerful: it can dictate when simple tip growth is to change to branching, and can also say something about what branching angle is to be achieved. This is the feature mentioned in

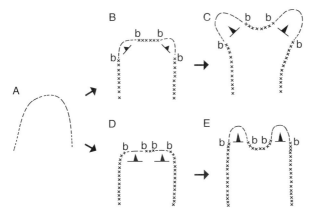

Figure 5.5 A theory for how branching angle may be controlled, used successfully in computations. Boundaries between active and 'dead' regions are marked: b.××× morphogentically 'dead' region; - - - low X concentration; — — — high X concentration. Faster upward advance of b boundaries pushes branches up onto the flattened tip, giving low-branching angles. From Holloway and Harrison (1999a), with permission.

the introduction to this chapter as pointing towards a timing mechanism inside the cell that can communicate with the surface events, and as being probably much more general than the specific Brusselator mechanism we have explored.

5.3 3-D COMPUTATIONS (BY L. G. HARRISON AND DAVID M. HOLLOWAY)

5.3.1 Going from 2-D to 3-D, and some results

Given Lacalli's opinion that the simplest computations are the most informative, together with my opinion (Harrison 1982, p. 28) that one of the valuable features of hierarchical linkage of events is to reduce the spatial dimensionality, e.g. formation of an annulus defines a 1-D domain for formation of a pattern of repeated parts, why should one go into the formidable labour of setting up programs for 3-D computations at all? For *Micrasterias*, there are at least three good reasons: first, in 2-D work all the branching and tip-growing processes are necessarily coplanar, since we are drawing shapes in a plane. It is a striking feature of *Micrasterias* morphogenesis that the great majority of these events are coplanar, when they happen in a 3-D world that

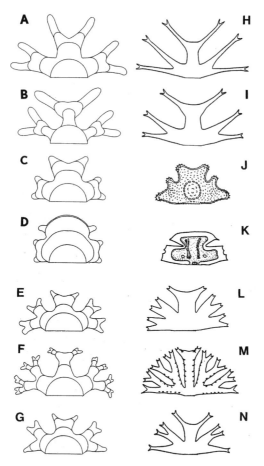

Figure 5.6 Right-hand column, H–N: for six species, varieties or variant forms of *Micrasterias* and one species of *Euastrum* (J), fully formed cell outlines, reproduced from Prescott *et al.* (1977) by permission of the University of Nebraska Press. © 1975 by the University of Nebraska Press. Left-hand column, A–G: computed cell outlines at four stages of development matching each observed shape. From Holloway and Harrison (1999a), with permission. See that reference for full details of changes in X_{th} and the timings of these changes required to achieve these species-specific matchings.

has no such restriction. What establishes this coplanarity? Second, a few species do make one out-of-plane branching, on the polar lobe (Fig. 5.7A). Third, the 2-D work seemed capable of giving most of the story for dome-shaped tips indulging in simple tip growth or branching events; but the clefts between growing lobes, i.e. saddle points, are a different matter (Fig. 5.7B). In the 2-D work, these points tended to be

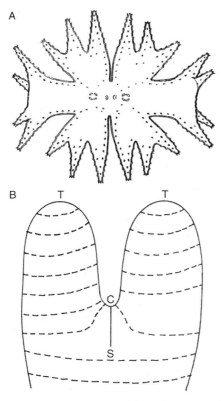

Figure 5.7 Features requiring 3-D computations. A: out-of-plane branching of polar lobes of *M. mahabuleshwarensis* var. *Wallichi*, from West and West (1905). B: sketch of tips T and cleft C, showing that the 2-D approximation (solid curve) may be expected to give a good account of growth near T but not near C. The solid line CS is intended to suggest a section through the cell surface in a plane perpendicular to that of the diagram. In this section the saddle point C would appear like a tip T, but controlled in upward advance by the growth of the remainder of the section below it.

moved outwards by the operation of the growth algorithm even when the growth catalyst concentration X had decayed to zero at those points; such movement was artificially prevented in the program. In a fully 3-D computation, such points would not move. The point C in Fig. 5.7B is the bottom of a cleft in the section in the plane of the paper, but the top of a non-growing region in the perpendicular plane. That non-growing region tethers the point C so that it cannot move upwards; it has no analogue in the 2-D work.

The 2-D growth algorithm illustrated in Fig. 5.3 uses an obvious and unique representation of a curve by a finite number of nodes and interconnecting lines. There is no corresponding unique representation of a curved surface in 3-D space. One of the most popular ways to represent the surface, which we have used, is a triangulation. In our work (Harrison *et al.* 2001) the initial shape has most usually been a hemisphere, approximated by nodes each joined by lines to six neighbours, forming an array of rather roughly equilateral triangles (Fig. 5.8B). In our reaction–diffusion and growth calculations, the six triangles around any one node are used together as a 'finite element'. As the surface grows, more triangles must be inserted in the rapidly growing regions to keep the sizes of the triangles and hence the precision of representation of the surface as uniform as possible and close to that of the starting shape. Fig. 5.8C, from Kaandorp (1994), is for a quite different growth process from the one we are studying, but illustrates well a dichotomously branching shape with ongoing refinement of the mesh by addition of small triangles in many places. If one is to avoid the formation of some very long thin triangles, and unequal representation of what is going on around various nodes by varying coordination numbers to neighbouring nodes, the method of adding new nodes, lines or triangles has non-trivial difficulties. We devised a method that limits coordination numbers to five or six (except on the original equator of the hemisphere). Following Kaandorp, we insert a new triangle (three new nodes) into an old triangle, with the three new vertices at midpoints of the old sides, as soon as any one side of the old triangle has exceeded twice its original length. The excessive length is discovered when one particular node is being visited in the computation. To avoid new nodes having a coordination number of four, a new triangle is inserted in all six of the triangles forming the finite element around the node being visited. In Fig. 5.9, node A had been surrounded by large triangles like those around node C, but a new triangle has been inserted in each of them, e.g. the one with B as one vertex. That vertex lies on the boundary between a refined region and a slightly slower-growing region in which no edge has yet reached the length demanding refinement. So: are B and C neighbours or not? If yes, C has seven neighbours; if no, B has four. We solved that problem by making them neighbours when B was being visited, but not when C was being visited.

We show this much detail of how the mesh has to be handled to give the reader the flavour of the kind of work that has to be done to implement an intrinsically rather simple growth algorithm in the computer, for 3-D growth. Much of the complexity in handling the mesh arises because we require it to respond to spatial symmetry-breaking events generated by the reaction–diffusion chemistry without imposing any constraints on the resulting shape changes. This kind of work may

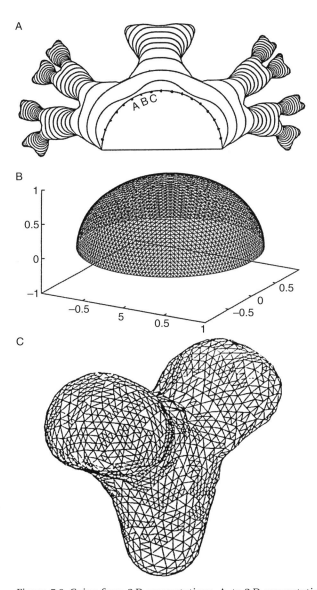

Figure 5.8 Going from 2-D computations, A, to 3-D computations, B and C. The initial semicircle in A, with three nodes labelled A, B and C, is like the general schematic of the 2-D growth algorithm in Fig. 5.3. In B, for the start of a 3-D computation, this is replaced by triangulation of a hemisphere. As the surface grows, the mesh must be refined from time to time by insertion of new triangles in fast-growing regions. C illustrates this with a figure from Kaandorp (1994), with permission.

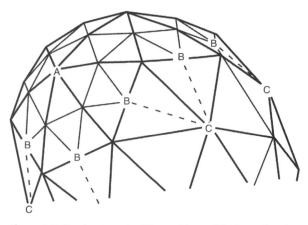

Figure 5.9 Our treatment of the problem of the boundary between refined regions and parts without new nodes. We allow a node to have only five or six neighbours. To achieve this for nodes B and C, when node B is visited in the computation it recognizes C as a neighbour, but when node C is visited it does not recognize B. From Harrison *et al.* (2001), with permission.

seem to be at a great remove from the work of geneticists and molecular biologists. But the substances A, B, X, Y and the enzymes governing the rate constants *a, b, c* and *d* must ultimately be identified as biochemicals related to the genes. If our hypothesis is correct that some such suite of substances patterns the distribution of a growth catalyst for surface area, then the handling of growing meshes in the computer is essential to study how a pattern derived from a number of genes is finally expressed morphologically.

In our 3-D program using the mesh as just described, there remain many possible ways to go about computing the reaction–diffusion dynamics and the outward movement of the surface in response to its increase in area. In the present work, we used methods based on averaging over each six-triangle finite element devised by Stephan Wehner, and described briefly in Harrison *et al.* (2001). Values of concentrations were defined at the nodes, but then averaged over the three nodes of each triangle and multiplied by its area. For diffusion, concentration gradients were approximated as constant from the outside edge of each triangle inwards to the central node of the finite element, and again triangle area was taken into account. Both reaction and diffusion were summed over all triangles of the finite element. For diffusion, the use of finite elements rather than finite differences along line segments gives better averaging to cope with distortions of the mesh during growth calculations.

The growth algorithm was necessarily quite different from the intersection-of-two-circles method used in our 2-D work. (If, instead of triangulation, we had used a honeycomb-like array of hexagons, as in Fig. 4.5, an intersection-of-three-spheres method could have been used; but the triangulated mesh is superior for most aspects of the calculations.) In 3-D, we used the entire finite element to compute the movement of its central node. The direction of movement was calculated as the average of the directions of the vectors normal to each of the triangles of the element. The amount of movement was calculated by finding the area increases for three test movements, fitting their results to a quadratic function, and using the function to calculate the movement for the required increase. This method works well for all parts of the surface that are convex outwards, but is still deficient in clefts. In all the work we have so far done, however, the catalyst concentration X is very low in such concavities, in fact always less than X_{th}, so that there has been extremely low growth in these regions.

Fig. 5.10 summarizes the kinds of results that have been obtained in 3-D computations (published in Holloway and Harrison 2008).

(A) Continuous elongation of a cylindrical stalk by tip growth. The somewhat fluctuating direction of the growth is essentially normal for a plant that is not plotting a compass course for its travel by any tropism for gravity or light.

(B) Tip flattening followed by resumption of apical advance. It has been a common feature of many of our computations, in 1-D or 3-D, that shift of growth action to widely separated side branches or to an annulus, as in this case, is usually followed by resumption of apical action; in reaction–diffusion mechanisms, this arises rather naturally without the need for additional features to force the model to do this (cf. Fig. 5.1F–G).

(C) Dichotomous branching.

(D) Formation of a whorl of three branches.

(E) Detail of (A), showing the mesh on which the computation is being done, which is distinctly more irregular than the morphology that it is being used to compute. We believe that the mesh is well-behaved in spite of developing such irregularities, first, because of the averaging over six-triangle finite elements that Wehner built into the reaction, diffusion and growth parts of the program, and, second, because of our introduction of a way of refining the mesh that preserves the coordination number of six.

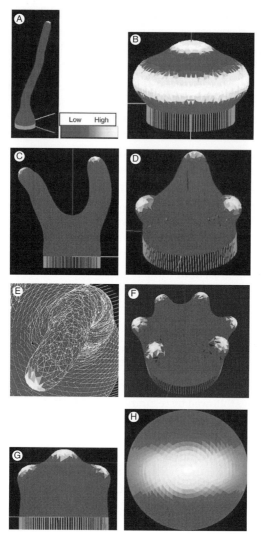

Figure 5.10 Results of some 3-D computations. Each computation starts from a hemispherical shape. Grey-scale is for the concentration of the Turing-patterned growth catalyst X, white high, black low. A: simple tip growth. B: tip flattening and resumption of apical tip growth. C: dichotomous branching. D: three-branch whorl. E: detail of A, showing the mesh used in the computation. F: a six-branch whorl achieved with a single reaction–diffusion patterning without prior annular pattern formation. G: a three-fold branching with all three branches coplanar, as happens at the beginning of *Micrasterias* morphogenesis; to achieve this needed two-stage patterning, the first stage being shown in H. Adapted from Holloway and Harrison (2008), with permission.

(F) A whorl of six branches, achieved by using a single
    Brusselator, not a two-stage model.
(G) A triple branching with all branches coplanar, not in a whorl.
    This is what is needed to explain how *Micrasterias* maintains
    its planarity. This one does use a two-stage model. On the
    original hemispherical surface, a previous patterning
    mechanism has drawn a stripe of high concentration of one
    of the reactants (A) for the second stage.
(H) This is the A stripe controlling the coplanarity of branches
    in Fig. 5.10G.

We tried two ways of controlling the morphogenesis to make
branches coplanar. The first was to vary the starting shape, using
a semi-ellipsoid instead of a hemisphere. This did not work (even
up to a 16:1 axial ratio). The second was to use a previous
chemical patterning stage as illustrated in (G) and (H). This was
much more powerful, and capable of doing the job. In studying
mechanisms for developmental patterning events, both the
powers and the limitations of hierarchical linkage of successive
events are important. Here, we show an instance where upstream
control is necessary. In Chapter 8, in relation to *Drosophila*
segmentation, it is indicated that downstream autonomy
may be an essential feature of the stage that generates the
segmental pattern.

5.3.2 Available software

For a long time, one of us (LGH) has had at the back of his mind an
ambition to turn a computer program on pattern formation, that had
been developed in a postdoctoral project, into a piece of software that
could be used by experimental biologists without the need to write any
software or to know how to handle the differential equations on which
it is based. (More commonly, the tens of thousands of lines of code that
one postdoc has written in Fortran or nowadays in C are perfectly
unintelligible to the next postdoc who comes along to continue the
work; and understanding the differential equations is prerequisite to
being able to write and handle the programs.) Eventually, the other of
us (DMH) got support from the British Columbia Institute of Technology
to turn the 3-D programs developed by Wehner and him into just such a
generally usable piece of software. This is now available for free down-
load from http://commons.bcit.ca/math/faculty/david_holloway/plant
growth/index.html, copyright D. M. Holloway.

Available, one may ask, to do what; just how can such software assist in bringing experiment and theory together in developmental biology? In Section 1.6, I referred to the use of 'pictorial reasoning', with the examples of shapes of atomic orbitals and their combinations into molecular orbitals being widely used by people who are not going to spend their time wrestling with Schrödinger equations. Earlier (Harrison 1993, Chap. 3), I devoted a complete chapter to 'pictorial reasoning in kinetic theory of pattern and form', using as examples spatially unequal mitotic events in plant cells and response to damage and grafting in animal development, including the reaction–diffusion interpretation of *Hydra* grafting data in the first modelling paper of Gierer and Meinhardt (1972). In this book, see also the quotation from Meinhardt (1984) in Section 6.3.5, ending 'we hope that our models can help to design appropriate experiments'. How, then, can the use of the 3-D plant growth software assist anyone in experimental design?

The essence of this kind of program is that it is a thing, residing in the computer, that will behave somewhat as a developing plant behaves because it contains mechanisms that we hope have some relationship to mechanisms in the real plant. Thus, although what one is doing with the computer is entirely theoretical work, it has the flavour of experiment, and indeed is often done best by experimental scientists. (Wehner, having come into this modelling work after a PhD in pure mathematics, was one day busily taking notes on what appeared on the computer screen while I was looking over his shoulder. He asked: 'Am I doing this right? I'm not used to writing down what I see.' He was having to behave like an experimentalist, or perhaps even like an observational biologist.)

The thing has a shape, upon all or a substantial part of which the chemical interactions of Eq. (5.1a–d) and Eq. 5.3 have generated a patterned distribution of the concentration X. These patterning inter-actions have much of the character of regulatory processes, both up- and down-regulations, but their dynamics are of such character that their effects cannot be understood by drawing a schematic of sub-stances regulating each others' production and just looking at it with some verbal reasoning. The mathematics is necessary (because the concentration dynamics depend on the quantity of a continuous vari-able) and the user has been spared having to do it by the computer program, which the user does not even have to see. The on-screen pictorial results will show what the interactions do. Some of it may be quite obvious, and some of it quite counter-intuitive.

One of the latter aspects is that a user may acquire a new perspec-tive on what parts of a complex pattern need more than one process to account for them, and what parts are produced together quite easily by a single process. For instance, in Sections 2.2 and 3.2, and Harrison and von Aderkas (2004), it was shown that, on the flattened apex of a conifer embryo, three features (the cotyledon primordia, their inset from the edge and the extensive flat central region) do not require consideration as separate regions where different processes are going on, but are all generated together by a single waveforming process. Just how much of a pattern can be one entity, and whether it can be made in a single patterning process or needs a hierarchical linkage of processes, can be explored with the aid of computations with this kind of software. This is potentially useful in thinking about the shoot apical meristem and its relation to production of lateral organs at the shoot apex, topics that tend to generate continuing controversy.[1] I am not directly involved in that controversy, and shall not attempt to review it here. The flavour of the exploration of this topic in terms of gene activities is well conveyed, for example, by Reddy and Meyerowitz (2005), especially in relation to a gene product (CLAVATA3) restricting its own domain of expression and the overall size of the shoot apical meristem in *Arabidopsis*.

The software lets the user explore many of the aspects that go into the interplay between geometry, chemistry and growth rate. Initial geometry, for instance, can be changed, to see the effect on chemical pattern. Initial patterns (such as in Fig. 5.10H) can be added, to explore the effect of hierarchical patterning. For given chemical parameters, the rate at which chemistry catalyses surface growth can be altered, to explore this crucial feedback cycle (since growth will also alter the chemical pattern). This feedback can also be altered by varying the X_{th}. With or without growth, the software also begins to give a feel for how reaction–diffusion makes pattern. For instance, by successively decreasing the diffusivities, one can decrease wavelength and see a progression to higher complexity, such as in Figs 5.10A–D.

To choose values for all these parameters starting without any guidance would require a good knowledge of the mathematics of Turing patterning. The intention of designing this software is that the user does not have to acquire that expertise. Starting from a set of demonstration values (for 'tip growth', 'dichotomous branching', etc.), the user can begin to slowly vary parameters, and build up a feeling for the

---

[1] See Smith *et al.* (2006), de Reuille *et al.* (2006), Jönsson *et al.* (2006), Hamant *et al.* (2008) for other recent modelling approaches.

dynamical interactions controlling patterning and shape. It is hoped that seeing how ranges of rates affect behaviour (for instance that increasing *A* or decreasing *B* can shut off patterning), can begin to replace simplistic concepts of up- and down-regulation – to build an appreciation for the dynamics inherent in such a relatively small system (compared, for example, to a full gene network). We feel it is this sort of exploration of a model that can lead to intuitions for new types of experiments to test the dynamical aspects of development, and therefore heartily promote its use by experimental biologists.

5.4 BACK TO 1-D: THE MOVING BOUNDARY; AND WHAT IS POLARIZATION?

In all the 2-D and 3-D work described in this chapter, the phenomena being modelled involve the maintenance of limited size for the morphogenetically active regions while they do the job of making the whole plant grow in size. So any mechanism we postulate within these regions has to have some kind of feedback into the control of its own boundaries. Holloway's introduction of the threshold concentration X_{th} below which the pattern-forming mechanism collapses, has proved very powerful for making a Brusselator mechanism control the boundaries within which it can produce pattern. But it must of course be an abbreviation, summarizing the result of an unstated chemical mechanism, designated 'feedback loop III' in Figs. 2.1C and 3.1. I wanted to make the postulated mechanism complete by including some chemical kinetics to explain X_{th}. The obvious aspect to look at for a Brusselator mechanism is the production of the reactants A and B from which morphogens X and Y are made. I looked first at straightforward feedbacks from X or Y to the formation of A or B, in the same spirit as the k_pSX term in Eq. 5.3. In brief, although such models incorporate A or B as an extra morphogen, they turned out to behave essentially just like the two-morphogen models, and did not supply the required extra feature.

Next, I wanted to look at a boundary control model based on the property of Brusselator systems that patterning happens only for certain ranges of concentrations *A* and *B*; e.g. for a constant *A*, there is a threshold value of *B* below which patterning ceases and the dynamics produce only spatially uniform concentrations. Use of this property replaces the unexplained X_{th} by a threshold *B* fully explained by the mathematics of Turing dynamics, as applied to the Brusselator mechanism (see Chapter 6, and Harrison 1993, Chap. 7 and Section 9.1). But this concept is only going to work to limit the size of the active patterning regions if B is somehow previously patterned, defining regions of high B and regions of low B. This led me towards a word, and the concept it expresses, very commonly used in developmental biology: 'polarization'.

In algae or fungi, development often starts from a more or less spherical cell, e.g. a zygote as in *Acetabularia* or the common brown alga *Fucus*; and fungal hyphae also start from such a cell. It is a common phenomenon that development starts with a cylindrical tip-growing outgrowth emerging from one side only of the cell, which is said to be 'polarized' when it can thus specify action to happen on one side only. Polarization is itself a patterning event, susceptible to explanation by, e.g. a reaction–diffusion mechanism.

Eventually, Richard Adams and I settled down to manipulating a more chemically complex model containing feedback loops I, II and III together with a previously patterned input that we called 'initial polarization'. To study this level of chemical complexity without problems of mesh control and mesh refinement, we went back to 1-D and used growth only moderately elongating the system, so we could use a 240-node line without insertion of extra nodes. The simplification to 1-D means that an initial spherical cell has been replaced by a circle, and our 1-D line of nodes represents distance around the perimeter of that circle, as in Harrison and Kolář (1988) or the semicircular starting shape in Holloway and Harrison (1999a), shown here in each of Fig. 5.6A–G.

If pattern formation is to be confined, in the first instance, to a part of the system, to test that it can thereafter maintain the restriction to that same spatial extent, then the computation has to begin with a specified spatial distribution of B. We used one cycle of a cosine curve (plus a constant to avoid negative concentrations) with a minimum at each end of the system and a maximum in the middle (Fig. 5.11A). This defines some length symmetrically placed in the middle of the system as the region with B high enough for pattern formation. This input is what we called the 'initial polarization'.

The model for tip growth is the two-stage linkage of two Brusselators as suggested in Figs. 2.1C and 3.1 (feedback loops I and II). For these we use the terminology A_1, B_1, X_1, Y_1 and A_2, B_2, X_2, Y_2. The B of the initial polarization and of the first stage is B_1. The catalyst for morphological growth (elongation, in 1-D) of the system is Y_2. A computation is structured as numerous short repeats of:

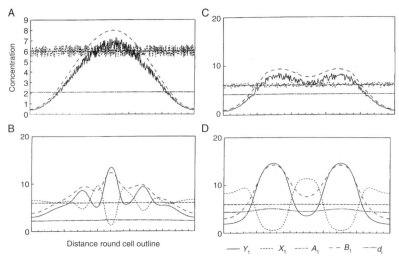

Figure 5.11 Ongoing computational research: exploration of degrees of downstream autonomy from initial polarization when the downstream dynamic elements are the three-feedback-loop scheme sketched in Figs. 2.1C and 3.1. In each pair of graphs, the upper (A or C) shows initial input patterns of A_1(constant, value 6) and B_1 (the 'initial polarization', a simple cosine curve with central maximum or a two-maximum curve made up out of cosine curves). For initial values of the Turing morphogens in a Brusselator model, $X_1 = A_1$ and $Y_1 = B_1/A_1$ (Harrison, 1993, p. 269), and some random concentration noise has been added to both of these. The lower graph (B or D) shows the patterns of the first-stage substances (subscript 1) and the internode distances d_i (Fig. 5.3) at an early stage in the computation (40 000 interations for each Brusselator). See text for the significance of these results.

1st stage R–D $\rightarrow$	feed forward $\rightarrow$	2nd stage R–D $\rightarrow$	Y_2–catalyzed growth $\rightarrow$	feedback to 1st.
A_1, B_1, X_1, Y_1	Y_1 to B_2	A_2, B_2, X_2, Y_2	all d_i s increase	Y_2 to B_1

In this model, 'feedback loop III' was composed of a feed forward in which B_2 is proportional to Y_1 and the feedback is achieved by adding a contribution to B_1 proportional to Y_2, so that total B_1 was the feedback plus the initial polarization. But the initial polarization was made to fade away by decaying exponentially with time. Thus the growing tip was obliged to become independently maintaining; in the end, it generated its own input B_1, without needing anything from the initial polarization. (In the event, we found that we had devised a positive feedback loop that was all too prone to blowups, and we had to limit each Y-to-B conversion by a saturation condition, i.e. making it proportional to $Y/(Y + \text{constant})$.)

For long, more or less straight extensions formed by tip growth, such as fungal hyphae, biologists usually extend the term 'polarization' to the directional control of the extension of the hypha, and refer to any change in this directionality as 'loss of polarization'. We are trying to examine what such terminology means in terms of putative chemical substances and their mechanisms of interaction. In the present model, the substance B_1 carries, in the spatial distribution of its concentration B_1, the directional information called polarization. But two different mechanisms are responsible for initially giving that information to the growing tip and for later maintaining it in the tip as it moves way from the original circular (or spherical) outline of the cell. The model therefore gives a more precise meaning to the relationship between 'polarization' as initially orienting the tip, 'polarization' as maintaining that orientation, and 'loss of polarization' as indicating any deviation in direction. And the model has taken a step towards supplying a chemical mechanism for the 'feedback loop III' necessary for understanding this morphogenesis.

A good mechanism (in which 'good' means 'probably related to how the plant is actually doing things') should display its virtue by addressing more questions than it was set up to address. Our main objective was to find a chemical-dynamic mechanism for the boundary of a growing tip to advance in lock-step with the rest of the tip. But once we had got this working for simple tip growth, we tried to tinker with parameter values to make the tip undergo an immediate dichotomous branching. And we found ourselves up against a difficulty: this model is quite reluctant to produce a minimum Y_1 right at the pole of the tip, which should be the saddle point between two new tips in a dichotomous branching event. The model has an observationally well-known property of plant growth: 'apical dominance'. And we didn't put anything into the model intended to do that. Fig. 5.11 illustrates this property. Graph A shows the initial polarization as the cosine curve of long dashes (B_1) or the solid line with noise added (Y_1). Graph B shows a result that was intended to be a simple dichotomous branching, but turned out to be two side branches with a dominant apex between them (the solid line, Y_1). Next, we tried to get rid of the dominant apex by somewhat flattening out the central region of the initial polarization. We did this by adding together cosine curves of different wavelengths (graph C, dashed B_1 line or solid Y_1). Graph D shows the result: a dichotomous branch indeed, but the separation of the two peaks has no resemblance to the chemical wavelength for the parameter values we put into the computation. The pattern is in fact a copycat version of the initial polarization, the two shallow peaks of which have led to dominance of two apices. Well, we know that apical dominance exists and we have inadvertently modelled it. But we also believe that there is a lot of downstream autonomy in patterning processes, and now we are trying to get it into this model. The work continues[2]. . .

[2] L. G. Harrison, R. J. Adams, D. M. Holloway, journal manuscript in preparation.

Part II   Between plants and animals

What am I putting between plants and animals? Physical chemistry, I think, because to me it is the filling that holds the sandwich together. I confidently expect that the theories of biological development will eventually embrace all kingdoms, just as, contrary to many expectations, genetic mechanisms proved to be so universal that what the plant-breeder, Gregor Mendel, discovered about peas has become the basis of the understanding of many human diseases, and much else.

Physical chemistry uses a lot of mathematical language, and it is a large part of my evangelistic attitude to suppose that much of developmental biology will some day have to be written in much the same language that physical chemists have been using for decades in their publications. But that is not yet accessible to many biologists. In this part of the book, I have had to struggle to say some things without using mathematical language. I remain unsure that I have done it, or that anybody could do it. In my previous book (Harrison 1993) I gave my reason for what I believe to be the necessary future adoption of much more mathematical language in experimental biology as follows:

> Mathematics is not essentially different from verbal explanation. Mathematical reasoning is simply the continuation of verbal logic by other means, when the complexity of the logic makes its expression in words cumbersome and obscure. (For instance, puzzles of the following kind are designed to exploit the equivalence between trivially simple algebra and quite obscure verbiage: 'Bill is twice as old as Joe was when Bill was ten years older than Joe is now; and Bill was thirteen years old when Joe was born. How old is Bill?') Occasionally, a view is put forward that mathematical reasoning is qualitatively different from verbal explanation and probably irrelevant to biology. I cannot argue against

that viewpoint, because I have never even begun to understand it. The essential equivalence of mathematical logic and verbal logic is to me an axiom, a credo. My book can cater to readers who are not fluent in mathematical languages, but it can do nothing for definite unbelievers in this credo.

It is not only a matter that mathematical and physical sciences versus biological sciences speak different languages; between physicists and chemists one can find a multitude of gaps. Consider, for instance, 'optical resolution', a topic I mentioned in Section 3.1.1. A sudden interest in it was the beginning of my move into developmental theory. It is a term in very common use in chemistry, when one is dealing with molecules that have both left-handed and right-handed forms, not superimposable by any rotation, as right and left hands are not. Chemical syntheses normally produce the two forms in equal numbers, making what is called a 'racemic' mixture. But living material normally contains only the form of one handedness, the other being completely absent. (Chemistry uses several precisely defined pairs of terms for the left- and right-handed forms: D and L, R and S, + and −. I shall use D and L, or d and l in the quotation below.) A process that disturbs the equal balance is called 'optical resolution', the word 'optical' getting in there because the two molecular forms rotate the plane of polarization of light passing through them in opposite senses, clockwise or anticlockwise. But I discovered that the term was quite puzzling to many physicists, to whom 'optical resolution' referred only to how small an object one could perceive the details of under a microscope, a rather different phenomenon but undoubtedly optical. The chemists' term is concerned with matter and molecules thereof, the physicists' with radiation and photons. My impression has been that to some physicists and mathematicians, my use of the word 'optical' in the chemical sense has been quite baffling, and has misled them seriously as to what I was talking about. When I asked physicists what they would call left- and right-handed forms of molecules, I was unable to elicit the reply 'enantiomers', a word that is as common to most chemists as 'dog' or 'cat'. The problem here was vocabulary; the concepts are well-known to physicists, but they would most probably like to talk about 'parity', while chemists like 'chirality', the Greek-derived synonym for 'handedness'.

Let us now contemplate 'non-linear dynamics'. Anything in science described by the term 'non-linear' can thereby be transported into the realm of mysticism. Algebraically, 'non-linear' means no more than that an expression doesn't just contain X and Y; it also contains X^2, Y^2 or XY, or terms in any power of X, Y or combinations of them other than the first. What does this do that is so generally important in the properties of anything in the physical universe?

This question can certainly invite obscure answers, even from people intending to be as interdisciplinarily clear as possible. But there are simple illustrations, and one of them is 'optical resolution'. Though I devote much less space and attention to this process in this book than in its predecessor, I start with it because I believe that it is a very quick and easy example to show the power of non-linear dynamics with no need even to write down one equation. This was done by Mills (1932) in a talk on stereochemistry at a regional meeting of the British Association. It seems to have clear priority for the concept of symmetry-breaking by an autocatalysis at a rate dependent on a higher power than the first of product concentration, and specifically for the loss of one enantiomeric form of any asymmetric molecule early in molecular evolution, so that now most living organisms have only L-aminoacids and D-sugars. I repeat the quotation from Mills I used in 1993; and I have to admit that it gives the lie to my contention above that the way out of obscure wording is into algebra. Mills did the job beautifully, in words. I regard it as an exception:

> It may be profitable to enquire whether the property of growth which is characteristic of living matter may not necessarily lead to its dissymmetry ... Let us now consider the growth of a tissue which is not completely optically inactive, that is, a tissue in which the d- and l-systems are not present in equal quantities. Let us suppose, for example, that there is twice as much of the d-system as of the l-system ... in the process of growth ... the d-system will increase at a relatively greater rate than the l-system. The complex dissymmetric components ... will be built up ... by chains of synthetic reactions, and the rate of formation of the end-products will be controlled by the velocity of the slowest link in the chains. If we consider a case in which, as must frequently happen, this slowest link is an interaction involving two dissymmetric molecules and ... assume that, as in a simple bimolecular reaction, the reaction velocity is proportional to the second power of the concentration, then the rate of formation of the d-component will be four times that of its enantiomorph. If this applied to every dissymmetric constituent of the new growth, then, whereas there was twice as much of the d- as of the l-system in the old tissue, there would be four times as much of the d- as of the l-system in the new growth.
>
> It will be clear that, even though the reactions of living matter may be less completely stereospecific than I have, for simplicity, assumed, and though the velocities ... may increase more slowly with the concentration than according to the second power, yet as long as they increase more

rapidly than according to the first power ... any excess of one system over the other in the old tissue will become greater in the new growth ... There is an *a priori* probability that an optically inactive growing tissue would be, as regards its optical inactivity, in a state of unstable equilibrium ... From this point of view the optical activity of living matter is an inevitable consequence of its property of growth.

D'Arcy Thompson (1917) was surely advocating a similar perspective for biological development in general by entitling his book *On Growth and Form*, and by most of its content. His book is often referred to by the date of its second edition (1942), or read as Bonner's (1961) abridgement. One can lose a historical perspective if one does not recognize the 1917 date of the first edition, in which Thompson wrote that 'the *things* which we see in the cell are less important than the *actions* which we recognize in the cell' (Thompson 1917, Chap. 4), long before the invention of the electron microscope or any idea of how to find the structure of a gene.

Following what I have quoted, Mills went on to discuss the initial bias that triggers the departure from the racemic state, and to show that the expected scatter of equal numbers of d and l molecules would be adequate to start the asymmetrization in something about comparable to the size of a living cell, 30 µm in diameter. With that choice, and with the repeated use of the words 'tissue', 'growth' and 'concentration' in the above quotation, Mills has focused attention more upon the macroscopic scale than the molecular, and also upon the dependence of rates upon concentrations as the essential part of a theory of spontaneous asymmetrization. His suggestion was a signpost towards a dynamic theory of development. I took the route by writing two papers on the concept that, in a system such as might have been present in the earliest days of biochemical evolution, the processes of optical resolution from initially racemic composition might have involved intermediate stages in which oppositely resolved regions, i.e. some d and some l regions, would be transiently patterned (Harrison 1973, 1974 and several sections on optical resolution in Harrison 1993). See also Frank (1953); I think he has priority for the concept of the dynamics involving an increase in size of resolved regions above the molecular and movement of interfaces between such regions as essential features of the dynamics.

There are many other ideas for how spontaneous optical resolution happened; see collections of papers in Walker (1979) and the *Journal of Molecular Evolution* from 1974; and for an account of universal handedness by a grand master of putting science in popular language, Gardner (1982). Some of these concepts involve chiral asymmetry being passed on to living chemistry after arising elsewhere, e.g. in a source of circularly polarized light or even in the well-established concept in nuclear physics of 'non-conservation of parity' in the weak interaction. Linkage of chiral asymmetry in life to that property of the entire universe is favoured, I think, by people who believe in complete determinism from the start in the things that science deals with, leaving no role for randomness. To those who adopt Mills' idea, the optical resolution that we now see in living material could have gone either way in early biochemical evolution, so that we could have had life as we know it except that it would be the mirror image of what we now have on Earth, and would have D-aminoacids and L-sugars. Wald (1957) quoted Einstein as saying of the dominance of one chirality over the other, 'it won in the fight'. This seems curiously at odds with Einstein's better-known statement: 'I do not believe that God plays dice.'

Part II explores the development of an understanding of biological pattern formation in terms of the relative rates of processes – the kinetic viewpoint of developmental biology.

6

The emergence of dynamic theories

6.1 FROM WIGGLESWORTH TO TURING

The orderings, and orderings of orderings, that go into the generation of a human body, or even of much less complex organisms, make any catalogue of catalogues or bibliography of bibliographies look like rather simplistic stuff. It is unsurprising that much enlightenment on what is happening is derived from what happens when something goes wrong. The usual major defects known as mutations have been of enormous value in genetics. Thus, rather confusingly for anyone approaching the field for the first time, many *Drosophila* genes are named negatively, for what happens when they are not working: *fushi tarazu*, Japanese for 'not enough segments', is working properly when the insect produces exactly enough segments; likewise for *hunchback, Krüppel* (German for cripple) and so forth. Also, experiments in developmental biology often involve transplanting pieces of tissue to places in an organism that Nature, with her zeal for self-organization, would not have thought of.

But if perfect orderings are the ideal, Nature sometimes seems to persist in 'going wrong' by keeping on doing something rather sloppily. A kind of sloppiness that is potentially informative is imperfect ordering of a pattern, in which numerous parts (usually 'spots' on a more or less flat surface) are neither randomly distributed nor marshalled into square or hexagonal ordering, but something in between. Examples are quite diverse: pores in the cell wall of a unicellular desmid alga (Lacalli and Harrison 1978b); wool follicles on Australian sheep (Claxton 1964); rod cells in the retina of a cat or monkey (Wässle and Riemann 1978), shown in Fig. 6.1; and bristles on the integument of some insects (Fig. 6.2). This last, as studied on the tropical blood-sucking bug *Rhodnius prolixus*, attracted the attention of Wigglesworth (1940).

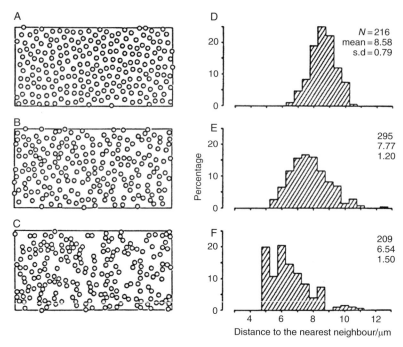

Figure 6.1 Partly ordered patterns I. Rod and cone cells in the retinas of mammalian eyes. A: rods in the cat; B: cones in the cat; C: rods in the rhesus monkey. From Wässle and Riemann (1978) with permission. D, E, F: the radial distribution functions for these three patterns.

He proposed that, around the initiation site of each bristle, there is a region in which the formation of additional initiation sites is inhibited. Indeed, for others of the examples mentioned, radial distribution functions (Fig. 6.1) show that the nature of the partial ordering is that the repeated structures keep apart beyond some minimum distance. Thus, small structures seem to be spaced out as if they were much larger. Some kind of action at a distance seems to be taking place.

The simplest possible mechanism for this is that some chemical substance needed for formation of a bristle primordium is diffusible, and thereby flows inward to each new primordium as it forms (Fig. 6.3). This depletes the surrounding region of the necessary substance to below the minimum level needed for primordium formation to start. An obvious alternative is that some chemical substance is produced by the primordium, flows outwards by diffusion into the surrounding region, and is active as an inhibitor of primordium initiation. For a picture of this, simply reverse the direction of the arrows in Fig. 6.3.

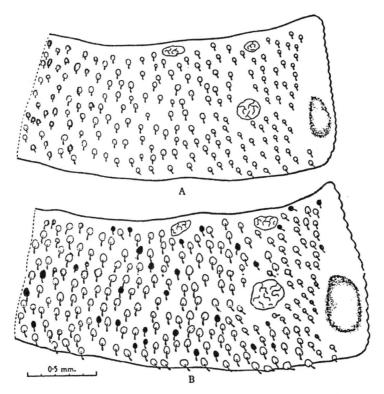

Figure 6.2 Partly ordered patterns II. Bristles on the integument of the insect *Rhodnius prolixus*, from Wigglesworth (1940). A: fourth instar; B: fifth instar, i.e. a moult, with a change in the pattern between A and B. The shaded bristles in B are the ones added in this change.

That reversal is not a trivial point. If, for both alternatives, depletion or inhibition of the surrounding region, the motion of the substance is simple diffusion, and the mechanism for allowing or preventing primordium formation leads to a simple threshold concentration value, then the two alternatives are equivalent in terms of mathematical description of their dynamics. Just as a matter of terminology, this means that the two alternative ideas can often be put together as one in discussing their dynamics; depletion can be thought of as a de facto inhibition by the loss of something needed for activation. Therefore, I shall sometimes refer to both together as activation–inhibition dynamics, avoiding tiresome repetition of 'inhibition or depletion'. There are times when we need to distinguish between the two, but not always.

Clearly, if we seek the detailed mechanism of the threshold concentration for switching on or off the possibility of new activation

Figure 6.3 Wigglesworth's (1940) sketch of his 'inhibitory field' mechanism. The arrows show movement towards each bristle primordium of a substance necessary for its formation, depleting a surrounding region so that no new bristles will form there. This is a de facto inhibition by depletion; for a true inhibition by a chemical inhibitor moving outwards, simply reverse the arrows.

at any given position, we may be led into some complicated biochemistry. And we have no a-priori indication as to whether the crucial events are likely to be close to molecular genetics (e.g. the depleted substance is a transcriptional regulator) or very far away from it (e.g. the inhibitor is some small molecule that blocks receptors on a cell surface, or inhibits an enzyme action). The possibilities are legion. Again, we see a set of phenomena with probable unity in dynamics and diversity in molecular chemistry.

The Wigglesworth concept is known as an 'inhibitory field' mechanism; and the two alternative forms of it are a 'depletion' mechanism and an 'inhibition' mechanism. From time to time, the concept has been discussed in relation to the original problem of partly ordered patterns (Claxton 1964; Lacalli and Harrison 1978b). But also, it has been suggested as a possible mechanism for locating the primordia of single organs such as the heart (e.g. in the axolotl, Sater and Jacobson 1990). For the latter, far more elaborate dynamic models have been devised, in the general category of mechanism that is capable of producing precisely ordered patterns: reaction–diffusion, of the kind first proposed by Turing (1952) and especially the form of this theory devised by Gierer and Meinhardt (1972) with particular attention to locating and localizing single structures. Holloway *et al.* (1994) applied the Gierer–Meinhardt theory to a substantial amount of experimental detail for the axolotl heart example (Section 9.1). Given that the Wigglesworth idea obviously contains both reaction and diffusion, is it in any way related to these apparently much more complex reaction–diffusion models?

There is in fact a close relationship, which is simple to state, but if followed up to make the Wigglesworth idea explain organ localization, soon demands the whole mathematical panoply of Turing and Gierer–Meinhardt reaction–diffusion. In the Wigglesworth concept, activators do not move. Activation events try to start in random locations and succeed except for regions where the event is suppressed by inhibition or depletion. Activator substances stay localized wherever they first appear, and need no further consideration in relation to geometry of pattern, except that they must lead to (e.g. catalyse) the formation of diffusible inhibitors by which their own growth is inhibited when they start to appear near an established activated spot.

If we had to write out all these processes as steps in a mechanism and consider all their rates, the theoretical treatment of the dynamics would already be quite complex. None of it has to be done so long as the activation events stay localized in their original randomly selected places. To be sure, some mathematics of the inhibitor diffusion is necessary to flesh out Wigglesworth's idea a little. Diffusion is a time-dependent process, and the inhibited regions would not magically appear instantaneously with a fixed size around each activated spot. Lacalli and Harrison (1978b) considered time-dependent diffusion as well as theories of the activation event of a similar kind to nucleation of crystallizations from solution, to try to correlate measures of degree of order in the partly ordered patterns with predictions from the theory.

But if the activator substance is also allowed to move by diffusion, everything changes in how one has to go about pursuing the theory. It becomes much more powerful, capable of generating fully ordered patterns. Activated spots may be destroyed or may change their positions so as to achieve this order. Patterns of chemical concentrations, as they emerge from uniformity, are at first harmonic waveforms. Wigglesworth's theory becomes Turing's reaction–diffusion theory, and what is going on cannot be understood without mathematical treatment of the dynamics.

It would be a very neat historical sequence if it could be said that Wigglesworth's idea gave inspiration to Turing, and Turing's theory formed the basis for Gierer and Meinhardt's. But that is non-history. Crystals may grow by a neat accretion of little pieces all locking into just the right places, but neither science nor living organisms grow that way. I know of no evidence that Turing took any idea from Wigglesworth in founding reaction–diffusion theory. Indeed, if we are to give credit where credit is due, the concept that wave equations for growth factors can be devised from autocatalysis or positive feedback loops is older than Turing's paper. Rashevsky (1940) proposed this concept while trying to advance theoretical biology as a respectable discipline at the University of Chicago. But his ideas did not embody the full panoply of activation–inhibition communication that is so powerful in the Turing concept. (It is interesting, however, that Rashevsky and Wigglesworth published their ideas in the same year.) Even two decades after Turing's paper, Gierer and Meinhardt (1972), though referencing Turing in their paper, in fact first devised their form of diffusible activator–inhibitor theory without knowledge of Turing's paper. Also, they referred to biological work pointing to the existence of gradients, but did not refer to Wigglesworth. History has more of Krespel than of crystals: a door here, a window there, all on whims of the moment; and finally, a house, that looks as if a plan had been executed.

An interesting aspect of this history is the citation record of Turing's 1952 paper (Fig. 6.4). It shows a striking increase from 1–2 citations per year up to the early 1960s, to about 100 per year in the late 1990s.[1] Unfortunately, the greater part of these citations are in the literature of the physical sciences. Even now, there are relatively few in the literature of developmental biology, which is the only discipline the theory was designed to serve. There is, at the moment, an opinion that Turing's theory sparked an initial surge of interest, followed by a long and deep decline, and that interest is now slowly resuming because mathematics is becoming more accepted in biology. That historical perspective doesn't exactly correlate with Fig. 6.4, does it? Among scientists *in general*, interest in the theory has grown slowly and steadily over 40 years, as the figure shows. What we still await is

---

[1] The citations went over 200 in 2009, following continued increasing growth through the 2000s. A survey of the citing articles is a fascinating study of the growth, maturation and application of the reaction–diffusion idea.

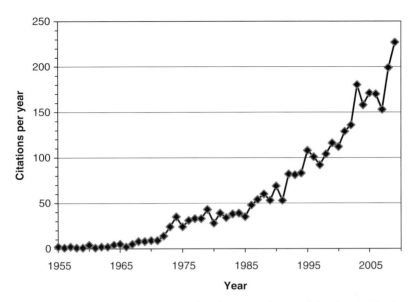

Figure 6.4 Citations per year of Turing's (1952) paper, 'The chemical basis of morphogenesis'.

a big meeting, almost a collision, but with more of a character of union than destruction, between the physical science of dynamics and the biological science of development. A graph of an apparently continuous function with some noise in it does not presage such events.

6.2 THE COMPONENTS OF TURING DYNAMIC PATTERN-FORMING MECHANISMS

The essence of Turing dynamics is that the theory shows how pattern can arise from a combination of activation, inhibition and communication across the region in which the pattern forms. It is by no means necessary that these processes should take the form of interactions of just three or four chemical substances and communication only by molecular diffusion. Activation here means the general property of something to produce, where it is located, more of itself out of other materials. This is identical to the property of assimilation, a fundamental property of all life. It can be thought of on different scales to the molecular, such as the cellular. Inhibition, likewise, can be on a variety of spatial scales, and communication is not necessarily only by molecular diffusion.

Nevertheless, the simplest way to approach Turing dynamics is to think of a pair of substances, activator X and inhibitor Y, the chemical kinetics of the reactions in which they are formed and destroyed, and the movement of both of them by molecular diffusion.

In my writings, I keep on plugging the concept of dynamic pattern formation in this way because I believe that it contains the general principle of how such pattern formation works in a way that should, to my mind, be in the fairly early education of all developmental biologists, just as the concepts of first- and second-order kinetics of reactions are known to all chemists, or Newton's laws of motion are known to all physicists. These things are all basics of how changes happen, applicable to a great diversity of substances and objects. The components of a Turing mechanism are the mathematical forms of the reaction rates and diffusion rates. Activation, inhibition and communication are processes and their rates, not things and their structures.

Because rates are described by rate equations, the statement in my previous book (Harrison 1993): 'All the words lead toward the equations, and the words are really useful only to people who are going to follow them that far' is definitely applicable to the topic of dynamic patterning. I have written the present section without equations (except for a chemical reaction equation) and tried to continue thus for the whole chapter; but at the beginning of Section 6.3, I give up. It can't be done, and in that section I try to use a minimal number of equations to explain why.

Let us suppose that two substances, X and Y, are uniformly spread across some region, i.e. at the same concentration everywhere. These substances are, in chemical-kinetic terms, reaction intermediates, formed out of reactant (substrate, in biochemical terminology) substances A and B and going on to be converted into products D and E. We are now at process II in Fig. 2.1C, if for the moment we remove the $_2$ from Y in that figure:

$$A + B \rightarrow X, \ Y \rightarrow D + E,$$

where in Fig. 2.1C X, Y is surrounded by curved arrows indicating Turing dynamic interactions between these two. Those arrows do not imply any one specific set of steps in a reaction mechanism. There are many types of reaction mechanism that have Turing dynamics, e.g. the Brusselator of Prigogine and Lefever (1968), as described earlier in Eq. (5.1), or a mechanism devised by Holloway et al. (1994), as given in Eq. (9.1), which gives dynamics close to those of Gierer and Meinhardt (1972).

Let us suppose that a long, narrow system (just so that we can think one-dimensionally in the first instance) has A and B supplied in a manner that keeps them continually controlled to constant spatially uniform concentrations. Turing dealt with growth of pattern out of uniformity, the biggest challenge to any explanation of pattern formation. Here, a question arises of whether it is at all necessary to do this. For instance, Gierer and Meinhardt (1972) devised their activation–inhibition–diffusion dynamics, having in mind a system that already contains a shallow concentration gradient of some precursor such as A or B, monotonic from end to end of the system. This approach, in effect, links the pattern-forming event under consideration to the result of a previous pattern-forming event. Biological development is replete with such linkages, and there is enormous opportunity for theoretical (especially computational) study of them, to try to match the details of observed phenomena. In this book, I have presented examples for *Acetabularia* whorls (annular pattern to whorl of repeating parts) and stellate *Micrasterias* cells (primary to secondary to tertiary dichotomous branches) in Chapters 3 and 5.

But to account for pattern arising out of uniformity remains both the fundamental challenge to pattern-forming theories and the easiest start into trying to understand the theories. So, from the spatially uniform start: in certain circumstances, having to do with the size of the region, the concentrations of the inputs A and B, and the quantitative details of rates of reactions and diffusive movements of X and Y, we may see the concentrations of X and Y become non-uniform, spatially patterned. Wait a minute; they won't. If the spatial uniformity is completely ideal, nothing will happen. The maxim that 'asymmetry begets asymmetry' is not completely invalid. But there doesn't have to be anything present like a long-range gradient. Any chemical system having some fluidity, however precisely set up on the macroscopic scale, will always contain some 'concentration noise'. This, looking overall quite disorderly, can be correctly regarded as being the sum of rudiments of spatially periodic patterns of a multitude of wavelengths, including the kind of spatial scale at which events in developmental biology usually first arise (tens of micrometres). All that is then needed for pattern formation is a selective amplifier for one or a small number of those rudiments. The domain where pattern is to form is a shop displaying a multitude of goods. The Organizer is the shopper and has one kind of item in mind. Selection is the name of the game. Turing said that one should not worry about how the pattern formation gets started, mentioning the analogy of an electrical oscillator. If it has been set up containing an LCR (inductance, capacitance, resistance) circuit such that it should oscillate at a particular frequency that can be calculated from the values of L, C and R, then you expect it to oscillate

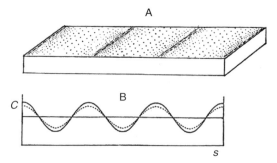

Figure 6.5 A: sketch of a rectangular tank containing a liquid solution. The shading shows the appearance of a coloured solute with a sine wave variation in concentration C along the horizontal direction. B: the sine wave distribution. This graph is referred to in the text with three different meanings for the solid and broken curves: (1) Diffusing solute, initially distributed as the solid curve; the broken curve is its distribution at a later time. (2) System with several solutes indulging in a pattern-forming mechanism; the broken curve is now the earlier time, the solid curve the later. (3) The patterns at one time for two solutes X (solid curve) and Y (broken curve) being the activator and inhibitor in a Turing pattern-forming mechanism.

at that frequency whenever you switch it on. You do not ask what little fluctuations in the circuit got it started.

A Turing reaction–diffusion mechanism is a selective amplifier for patterns. Let us consider particularly patterns of a number of repeated parts with a regular spatial periodicity. What does the mechanism find in the initial noise that gets it started making a pattern with specific spacing? The essence of Turing dynamics is that any harmonic waveform of spatial concentration distribution of the substances X and Y will grow or decay in amplitude in a manner dependent upon its wavelength λ, for 1-D pattern, or upon an equivalent to wavelength in 2-D or 3-D, e.g. $2\pi r / [l(l+1)]^{1/2}$ for a spherical surface harmonic of index l on a spherical shell of radius r, as introduced in Section 4.4. Let us look further at the 1-D case. The available rudiment can be described as a simple sine wave such as that in Fig. 6.5B.[2] The model starting situation here is a rectangular tank containing a liquid with a coloured solute dissolved in it. A striped pattern appears, Fig. 6.5A, because the

---

[2] Any pattern in 1-D can be described as a combination of sines and cosines (Fourier series) that will act as the single sine example here. In 2- and 3-D, the growth dynamics of periodic harmonics are analogous to the 1-D case described here.

concentration C of the solute varies sinusoidally with the long distance s of the tank. How will this system develop?

For our first look at this, let us suppose that there is no chemistry going on, nothing but the physical process of diffusion. This is a spreading process, heading for eventual uniform distribution of the solute. If it started with other initial states, such as a small blob of colour in the middle of the tank, one would appreciate spreading quite directly. But if one applies the mathematics of diffusion to the sine wave distribution, one finds that the sine wave will stay a sine wave, unchanged in wavelength (whatever value that has) and will simply decrease in amplitude A until it fades away altogether without ever changing in shape, as a pattern. The solid line in Fig. 6.5B, then, represents the pattern at one time, the broken line the pattern at a later time. In my previous book (Harrison 1993), I described the sine wave as the Cheshire Cat of patterns (referring to the property of that mythical beast that it faded away until only the grin was left, i.e. its features gradually lost contrast but not form). The sine wave pattern is something that diffusion cannot distort in form.

6.3 NO AVOIDING EQUATIONS

6.3.1 Diffusion: the need to think of curvature of a gradient

I tried to write this chapter without equations, and at this point found it impossible. It is exactly here that the topic becomes mathematically difficult for people who are not accustomed to using either elementary calculus or the diffusion equation as an everyday language. I hope that the reason for this becomes clear by the time I get to Eq. (6.2) and Fig. 6.6D. Why should a sine wave pattern have the peculiar property described in the preceding section? To consider this, it is simplest to discuss the sine curve in terms of concentration above or below the midline, which is the uniform value C_o to which C would fall when the grin of the cat finally disappeared. I shall use $U = C - C_o$. The sine curve is then:

$$U = A\sin(2\pi s/\lambda) \tag{6.1}$$

where λ is the wavelength and A the amplitude, i.e. the value of U at a peak of the pattern. Diffusion is flow down a concentration gradient, ideally at a rate proportional to the gradient, which is $\partial U/\partial s$. (To readers unfamiliar with partial derivatives and the writing of them with the symbol ∂ in place of d: when a function such as U depends

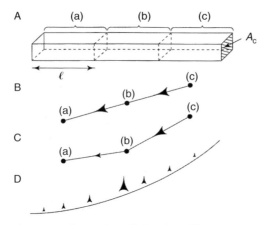

Figure 6.6 Illustration of why the diffusion equation (Eq. (6.2)) shows rise or fall of concentration as depending on the *second* derivative of concentration with respect to distance, therefore on the *curvature* of the concentration gradient. In B, C and D, the vertical axis is concentration. A, B and C refer to a three-box excerpt (boxes (a), (b) and (c)) from a finite-difference approximation, such as is used in computations. D is a possible continuum system for which C is the approximation. Arrows on B and C are rates of flow. Arrows on D are rates of rise of concentration, with arrow size proportional to rate.

on more than one variable, in this case position and time, $U = U(s, t)$, one most usually needs to look at these variations one at a time, and consider the slope of U versus s when t is constant and the rate of change of U with t at fixed position s. These derivatives are written with the curly ∂, as $\partial U/\partial s$ and $\partial U/\partial t$ and are called partial derivatives.) To anyone not previously familiar with the topic of diffusion, it could appear peculiar that the diffusion equation for rate of change of concentration anywhere along the system contains a second derivative:

$$\partial U/\partial t = D\partial^2 U/\partial s^2. \tag{6.2}$$

Here, D is the diffusivity of the solute (of concentration C, or its departure from uniformity U). The reason for the second derivative to appear is quite straightforward and is shown in a finite-difference analogue in Fig. 6.6. Consider a small part of the length of a 1-D diffusion system to be represented by three rectangular boxes (a), (b) and (c), as shown in Fig. 6.6A, with first-order exchange of the diffusible material between their end walls; the contents of each box are well stirred to constant concentrations. (This could be a little piece of the usual finite-difference approximation for a diffusion process in a computation.) Suppose now that the concentrations in (a) and (c) are held constant, so that we are interested only in how concentration in box (b) changes as time goes on. If the gradients are as in Fig. 6.6B, with the same slope from (c) to (b) as from (b) to (a), then the delivery of material from (c) to (b) is the same as the loss from (b) to (a), and the concentration in box (b) is constant. There is a concentration gradient, and there is movement down it by diffusion, but $\partial U/\partial t$ is zero at position (b). But if, as in Fig. 6.6C, the slope from (c) to (b) is numerically greater than that from (b) to (a), then more material is delivered to (b) than is removed from it in any given time, and the concentration at (b) is rising. What matters is not the gradient, but the difference between the (c–b) and the (b–a) gradient, i.e. how the gradient is changing with position. Converted from finite-difference to continuous variation of concentration with position, as in Fig. 6.6dD, this means that the rate of increase of concentration U at (b) depends on $\partial^2 U/\partial s^2$, as in Eq. (6.2). In Harrison (1993, Chap. 6), I pointed out that laws of motion containing only first derivatives, such as Aristotle's law that velocity (first derivative of position with respect to time) is proportional to applied force, did not urgently demand the invention of the differential calculus for their use; but laws such as Newton's, involving acceleration, i.e. the second derivative, did require this new mathematical language. That was in the seventeenth century. The language of the calculus is no longer new; in elementary form, which is all we need for the present purpose, it should be common knowledge to all scientists. The late Paul Green thoroughly understood this – see, e.g. the quotation in Section 1.4 and the title of Green (1999), containing the phrase 'combining molecular and calculus-based paradigms'.

We are now at the point where we can appreciate the nature of the problem of taking the next step when one has found good reason for suspecting that both an activator and an inhibitor are at work in a system and are distributed in spatial gradients. Fig. 6.6D and Eq. (6.2) show that what determines how diffusive transport is changing their concentrations is the *curvature* of each gradient. This is not readily amenable to the thinking in words that may have been very effective in getting to this point. It is, in fact, the great obstacle, perhaps the principal reason why there is such a persistent gulf between theory and experiment in developmental biology. The great jump into using calculus fairly intensively happens right here. I give, therefore, no apology for using a few more equations. This is not, however, a complete presentation of any way of solving the Turing equations. That is available in standard mathematical biology texts (e.g. Edelstein-Keshet 1988;

Murray 1989). For people (like myself) who have not taken formal courses in modern methods of solving partial differential equations, but who are quite familiar with elementary calculus, I devised the following curious amateurish approach to the Turing equations (Harrison 1993, Chap. 7).

6.3.2 Bringing in the reaction rates

However much one discusses diffusion, and considers Turing patterning, as many mathematically minded people do, as a 'diffusion-driven instability', nevertheless there are many scientists who are not going to acquire a 'gut feeling' for how Turing patterning works without tying it primarily to substances (or molecules) and how they interact with each other, with diffusion playing an ancillary role. Since reaction–diffusion goes nowhere at all without both reaction and diffusion, it is difficult to say which should be regarded as master and which as servant. In what follows, I want to bring in reaction rates, how they determine time dependence of the pattern, and how it eventually departs from simple harmonic shapes and also reaches a steady-state. Let us first, however, go back to the sine function, in order to get rid of it by dressing it in a servant's uniform. Its second derivative is proportional to itself:

$$\partial^2(\sin(\omega s))/\partial s^2 = -\omega^2\sin(\omega s) \tag{6.3}$$

where ω is inversely related to the wavelength, $\omega = 2\pi/\lambda$. This equation means that, wherever in our equations we see the second derivative of U, we can get rid of it and replace it with U multiplied by a constant which is wavelength-dependent. Since U is simply the departure from midline at any point along a sine curve, all such departures grow or decay, fractionally, together, so that the shape of the curve is not distorted. All the Cheshire Cat's friends will continue to recognize its face as it becomes more distinct or fades. This is the essence of how diffusion works in pattern formation. It implies also that wherever diffusion appears in a dynamic equation, it will be as a constant times a U or V concentration, looking just like a reaction term except that the constant is wavelength-dependent. That is the thing that gives the servants the keys to the castle.

For our second interpretation of Fig. 6.5B, let us suppose that we have several solutes in the system, with chemical reactions going on between them. It is then possible for the sine wave pattern, still without distortion of form, to grow in amplitude, i.e. in Fig. 6.5B the broken line would be the earlier time, the solid line the later one. This needs quite a special kind of system of chemical reactions. In the upper half-waves, concentration C is rising; in the lower half-waves, it is falling; and the same reactions are going on everywhere. The essence of Turing's idea was to devise the special chemical dynamics that could behave in such a way. As described in Section 6.1, these can be envisaged as an extension of the Wigglesworth inhibitory field concept, with additional features. There must be at least two interacting substances, X and Y. (Terminology: chemists will commonly denote a substance as X and its concentration as [X]. The latter gets very cumbersome in complex kinetic equations. I use the widely accepted way of getting around this. The name of a substance is Roman X, and its concentration is italic X.)

So, to add further confusion, I give a third interpretation for Fig. 6.5B: the solid line is the X pattern and the broken line is the Y pattern, both at the same instant in time. This is what a Turing pattern settles down to, with both patterns growing together, keeping the same ratio of their amplitudes. Two Cheshire Cats, X-cat and Y-cat, but both emerging from the grin into well-defined cats; and the sinusoidal form of the pattern is strongly linked to the fact that both X and Y are diffusible. Y diffuses faster than X. That point at least is fairly obvious from the Wigglesworth idea, and it is crucial. No two-substance reaction–diffusion mechanism acts to form pattern without it. But let us come back to Wigglesworth's inhibitory field concept, and contemplate Fig. 6.5B in relation to it. The peaks of the X distribution (solid curve) represent the activated points, e.g. where the bristles are going to form on the insect integument. The troughs are the inhibited regions; but they have now become exactly the same shape as the activated regions, upside-down. It is the diffusibility of the activator that has brought about this regularity, the sine wave form, and replaced the concept of a range of inhibition with that of a wavelength.

As to the chemical interactions of X and Y: the activator X encourages its own formation, but in a way that can somehow be interfered by Y ('somehow' means that I am here giving generalized features of a multitude of possible chemical mechanisms). If its self-activation has thus had the brakes put on it by Y, then X decays. That's what is happening in the troughs of the pattern, so that they advance downwards in just the same way that the peaks advance upwards. And it is quite evident from the relationship to Wigglesworth's inhibitory fields that X encourages the formation of the inhibitor Y, which diffuses out from activated regions.

Clearly it would be useful to write these interactions as equations; but there is a snag. In discussing a general concentration C,

I found it useful to use the measure U of departure of C from the spatially uniform value C_o. This departure can be upwards or downwards, so that while complete chemical concentrations must be positive, U can be positive or negative, and hence can be represented by a sine function without the need to add anything. Now that we are discussing two substances X and Y, it would be convenient to write $U = X - X_o$ and $V = Y - Y_o$. This I shall do. But what is going on if I write reaction rates as proportional to U and V, rather than to X and Y? Reaction rates commonly depend on whole concentrations, not differences. If we write them that way, it is possible to substitute U and V for X and Y, and see what happens to the rate equations. If we then hunt for leading terms in U and V, rather than in squares, products or any other functions of the concentrations, we are going through the process of 'linearization about the spatially homogeneous steady state'. If we do this and enforce an initial sine wave pattern, so that the diffusion terms change from second derivatives to terms proportional to U and V, we get:

$$\partial U / \partial t = k_1 U + k_2 V - \omega^2 D_X U \qquad (6.4a)$$

$$\partial V / \partial t = k_3 U + k_4 V - \omega^2 D_Y V. \qquad (6.4b)$$

The witchcraft that has been wrought by specifying the initial pattern as a sine wave is that the diffusion terms have changed into Cheshire Cats, and the distance variable has disappeared from them. The differential equations are now in terms of time dependence only, and hence are ordinary differential equations (ODEs, to mathematicians), not partial differential equations (PDEs). I could therefore have written dU/dt instead of $\partial U / \partial t$, etc. And they have begun to look so similar to the equation $dC/dt = kC$ that one might begin to expect concentrations varying exponentially with time as their solutions. I take that up again in Section 6.3.3.

Eq. (6.4a) and (6.4b) are the Turing equations describing what is going to happen to U and V if they are both made to start as sine wave patterns with the same wavelength $\lambda = 2\pi / \omega$. Just what, however, are those four rate constants k_1, k_2, k_3 and k_4? For instance, in one putative mechanism that has Turing kinetics, the Brusselator, given as a chemical mechanism in Eq. (5.1a–d), we had four rate constants a, b, c and d. Are those k_1, k_2, k_3 and k_4? The bad news is that they aren't. The ks are combinations of a, b, c and d and the two Brusselator reactant concentrations A and B; and while a–d must be positive, some of the ks can be negative. In fact, the nature of V as an inhibitor usually turns up as a negative value for k_2, so that more V means slower growth rate of U. All of this means that, short of sweating through the algebra or looking up these fairly complicated combinations in a table (e.g. Harrison 1993, p. 272), you can't relate the kinetics of pattern development at all simply to steps in a chemical mechanism. It can even turn out that the rate constant for the apparent autocatalytic growth of U, i.e. k_1, is related to decay steps in the mechanism. For the Brusselator, for instance, the algebra leads to $k_1 = Bb - d$. The rate constant of the auto-catalytic step, c, doesn't appear at all in this expression; and the things that do appear are from the steps in which X is destroyed, not the ones in which it is produced. Hence, to relate what is seen to be going on in pattern formation to the action of an activator or an inhibitor involves very indirect reasoning, commonly demanding computations. If one is trying to match theory to experimental details, a fairly close collaboration between two groups is often necessary. I give an example in Section 9.1, in regard to collaboration between my group and that of John Armstrong on heart formation in a salamander. Such people as Meinhardt and I, who keep warning people of counter-intuitive properties in dynamic pattern formation, aren't fooling. Pattern growth and chemical mechanism are related to each other in devious ways. There is no escape from the equations, because they and they alone tell you everything.

6.3.3 Exponential growth and pattern wavelength

Readers unfamiliar with differential equations may be feeling increasingly unenlightened up to this point; Eq. (6.4) does not really give you the 'gut feeling' for why pattern forms. Let us look at Eq. (6.4a) a little differently. Suppose that a pattern of sine waves like Fig. 6.5B exists, with the solid line as U and the dotted line as V. Let $\theta = V/U$, which is a constant all along the system. But the amplitude ratio θ could change with time. Let us suppose that θ settles down to a constant value on a time-scale shorter than growth or decay of the individual amplitudes themselves. You have to take my word for it that this happens. We can then look at change, say, in U by using θ to get rid of V and then writing Eq. (6.4a) as:

$$\partial U/\partial t = (k_1 + k_2\theta - \omega^2 D_x)U, \tag{6.4c}$$

where we are going to regard θ as a time-independent constant. In that case, we now have an expression showing that U at any position or its amplitude A will change exponentially with time according to:

$$A = A_o\exp(k_g t), \tag{6.5a}$$

where

$$k_g = k_1 + k_2\theta - \omega^2 D_x. \tag{6.5b}$$

The whole secret of pattern formation by reaction–diffusion lies in what's in this rate constant, which I call k_g to denote 'growth constant'. In fact, it can be positive or negative and therefore represent growth or decay of the pattern. (There is more. Much of the algebra that I am not showing here involves quadratic equations, and hence can lead to taking square roots of negative quantities. This leads into algebra of complex numbers, and a complex k_g means oscillatory behaviour rather than formation of continuously growing or decaying pattern.) Thus far, we have been looking at the wave pattern without regard for how long the waves are, in the pattern that finally gets established. The essence of the ability to form pattern lies in the dependence on wavelength λ (recall that $\omega = 2\pi/\lambda$). Obviously the last term in Eq. (6.5b) depends on wavelength and on a diffusivity; but so does the $k_2\theta$ term, in a more complicated manner, and with diffusivities of both X and Y in it (e.g. Harrison 1993, pp. 229–33). If the reaction mechanism has appropriate dynamics, which must include non-linear terms,[3] then for some ranges of values of the parameters in the dynamic equations, k_g is positive, representing growth of pattern amplitude; and that is only for a small range of wavelengths (Fig. 6.7, in part repeating Fig. 2.6). This is the 'band-pass' property of a Turing k_g–λ curve.

But in any given situation in a living organism, most wavelengths in the band give patterns that are a misfit to the size of the pattern-forming region. There are 'boundary conditions' to which the solutions of the equations are very sensitive, a common feature of partial differential equations. Consider the tank shown in Fig. 6.5 as the pattern-forming region. The pattern shown has three wavelengths exactly fitting the length L of the tank: $L = 3\lambda$. This is a good fit for what is known as a 'no-flux' boundary at the ends of the tank. There is no diffusion across the walls, and diffusion rate depends on concentration gradient, so there must be no gradient at the ends. That is what is shown. The sine curve is horizontal where it runs into the ends of the tank. To meet such a condition, the horizontal ends of the pattern need not both be maxima; one of them could be a minimum. The next possible good fits,

---

[3] A chemical mechanism with Turing's conditions of autocatalysis, inhibition and cross-interaction will produce U, V terms with powers greater than 1. E.g. Eq. (5.1c) for the Brusselator mechanism produces the non-linear terms UV, U^2 and U^2V. See also Harrison (1993), Sections 8.2.1 and 9.1.2.

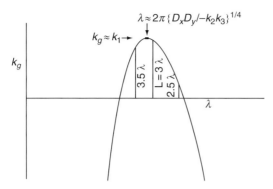

Figure 6.7 Exponential growth rate k_g of pattern amplitude versus wavelength λ for a particular choice of parameter values in a Turing mechanism giving good patterning behaviour. L is the length of the box in Fig. 6.5. The line $L = 3\lambda$ shows the growth rate for the three-wavelength pattern shown in Fig. 6.5, with meaning (3) for the two curves. Patterns with either 2.5 or 3.5 wavelengths fitting the same length L would have positive growth rates. See also Fig. 2.6 for a comparison of patterns that have positive k_g (growth) or negative k_g (decay).

for longer or shorter wavelengths, are $L = 2.5\lambda$ or $L = 3.5\lambda$. A possible set of k_g values for these three fits is shown as three vertical lines in Fig. 6.7. If one recognizes that the growth is exponential, so that a small increase in k_g is in fact a huge advantage, then it is clear that the 3λ pattern shown in Fig. 6.5 is soon going to be established exclusively of the other two that have positive k_g and zero gradients at the ends of the system. Fig. 6.7 gives approximate expressions for the optimal growth rate and wavelength at the top point of the curve. I give no further explanation of these. To get such expressions, one just slogs through lots of straightforward and tedious algebra. I've given all the strategy above.[4]

For the complete growth of pattern starting from uniform distribution of substances, plus a little 'concentration noise', that latter provides everything out of which pattern is to grow, and should be envisaged as having a rudiment of any wavelength that will fit the boundary conditions, corresponding to the three vertical lines shown

[4] In Fig. 6.7, the expression for k_g is a truncation of Eq. (6.5b). The full Eq. (6.5b) gives the relation between k_g and λ (through $\omega = 2\pi/\lambda$); the maximum in this relation, which is the fastest-growing wavelength, can be found by differentiation; the full equation for this 'chemical wavelength' is given in Eq. (7.22) of Harrison (1993); the expression in Fig. 6.7 is an approximation of this.

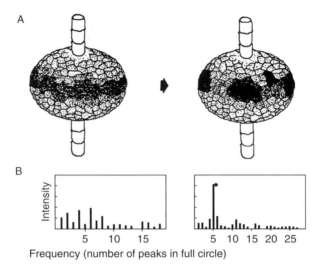

Figure 6.8 Whorl formation round the equator of a whorl mass in the slime mould *Polysphondylium pallidum*, from Byrne and Cox (1987) with permission. A: dark shading shows the whorl prepattern as revealed by anti-Pg101 immunofluorescent staining at two stages in whorl development. B: for samples at approximately the same stages, the power spectra of stain intensity. Each bar shows the intensity of a sine wave component of the pattern. Numbering on the horizontal scale is the frequency of the pattern as number of peaks in the full circle. The change in pattern towards growth of the five-peak pattern and decay of all others is the expected behaviour from Turing dynamics.

in Fig. 6.7 and many others outside the 'band-pass' region. For a coloured substance (or coloured light), a spectrum of intensity versus wavelength (or spatial frequency) would show many small peaks. As pattern develops, most of these would disappear, but one would grow (a pure colour would dominate). The only instance of experimental observation of this in development that I know of was reported from the E. C. Cox group (Byrne and Cox 1986, 1987; McNally and Cox 1989) for the cellular slime mould *Polysphondylium pallidum*. This species has a more complicated morphogenesis than the better-known *Dictyostelium discoideum* (for which see Section 9.4.1 and Fig. 9.12). *P. pallidum* forms a stalk with several 'whorl masses' spaced along its length; from each of these, a number of secondary stalks grow out sideways. The Cox group made several antibodies, one of which (anti-Pg101) detected a pattern of antigen expression on the whorl masses which appeared to develop into the pattern of points of emergence of secondary stalks. Fig. 6.8 shows the appearance of this staining at two stages, and the corresponding power spectra of stain intensity versus frequency. The spectrum develops according to the expectation I described above. In McNally and Cox (1989), they matched the development quite well with reaction–diffusion calculations. The antigen has never been chemically identified.

This may be a lucky instance in which the progression from random noise to definite pattern is slow enough to be picked up experimentally. Anyone who has experience of reaction–diffusion computations knows that this progression is usually very fast. When a pattern has become established and later changes to another pattern, that happens much more slowly (e.g. Fig. 6 of Harrison *et al.* 1981; Fig. 9.8; and the cover picture of Harrison 1993).

6.3.4 From exponential growth to steady-state pattern

The exponential growth of pattern amplitude clearly must become ridiculous after a while, as exponential growth always does. The mechanism (including boundary conditions) works to select a wavelength somewhere near to, but not exactly at, the maximum of the curve. Exponential growth is only an approximation, which is good while the pattern is just emerging out of uniformity, i.e. its amplitude A is still quite small. Beyond that stage, non-linear terms in the dynamics that have been ignored in the linear approximation begin to become important. For instance, the Brusselator model as given in Eq. (5.1) has terms in UV, U^2 and U^2V to be added to Eq. (6.4). The non-linear terms are not the same for all mechanisms that have Turing behaviour while A is small. But for any putative mechanism that is at all worthy of consideration as having anything to do with biological development, they all have one feature in common: they put the brakes on exponential growth. This entails also some distortion of the pattern from the sine waveform, usually different in the peak and trough half-waves. Fig. 6.9 illustrates this. It is a sketch, not a precise result of computation. Fig. 6.9A shows four time steps: the initial random input that gets things started; two stages (dotted lines) in the exponential growth phase (called 'primary patterning' in Fig. 6.9B), and the final steady-state at long (infinite) time. The Ms are overall concentration range, the same as $2A$ (twice the amplitude, top to bottom of sine curve) for the exponential phase. In Fig. 6.9B, M is plotted against time on a logarithmic scale, which converts exponential growth to a straight line. The distortion of the pattern from a simple sine wave that accompanies slowing and eventual stopping of the growth is called 'secondary patterning' in Fig. 6.9B.

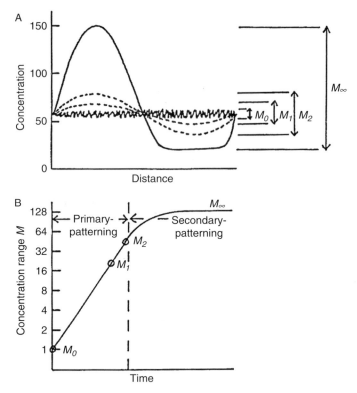

Figure 6.9 Progression of patterning from exponential growth to steady-state. A: sketch of a concentration pattern from random input to two stages of growth of sine wave (broken lines) to final pattern of altered form (but still the same wavelength). The way the form changes is different for different non-linear dynamics; the form shown is typical of Brusselator patterns. M is peak-to-trough concentration range. B: M (on a logarithmic scale) versus time, showing exponentially growing pattern (primary) and growth slowing to steady-state (secondary).

Turing-type dynamics are, however, potentially applicable to a great variety of patterns, and they are not all best thought about in exactly the same way. Diversity in patterning behaviour is often brought about by diversity in the non-linear terms that are not shown in Eq. 6.4. Many cases of this will demand the participation of theoreticians in ongoing biological work. But the properties of some of the better-known types of non-linearity should be well-known enough that experimentalists can use them in their thinking without such assistance, e.g. the contrasted properties of the Brusselator and the original Gierer–Meinhardt model.

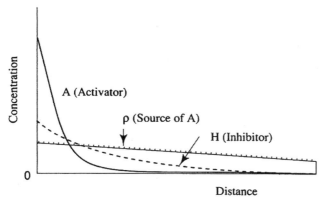

Figure 6.10 Steady-state distributions of activator and inhibitor produced along an elongated system by operation of the Gierer–Meinhardt dynamics upon a shallow source gradient. From Gierer and Meinhardt (1972), with permission from Springer-Verlag. See also the application of this mechanism to localization of a salamander heart described in Section 9.1, especially Fig. 9.2.

My own work, the greater part of it on plants as described in Part I, has been largely concerned with patterns of repeated parts and quantitative aspects of the spacing between adjacent parts. Here, the concept of wavelength is essential and one keeps on using formulae for the wavelength, e.g. that given for the Brusselator in Eq. 5.2. If, on the other hand, one is interested in the formation of single organs that are not repeated at regular intervals, e.g. a heart, as discussed for the salamander in Section 9.1, one is concerned with a phenomenon commonly known by the slogan 'short-range activation; long-range inhibition'. This clearly has nothing much to do with Fig. 6.5B, in which activator and inhibitor patterns go in lock-step right across the patterned region. But the slogan has everything to do with one of the commonest types of pattern generated by a Gierer–Meinhardt reaction–diffusion mechanism (Fig. 6.10). This has dynamics in the Turing category for primary patterning as defined in Fig. 6.9, but has different non-linear terms than the Brusselator, that give it very different secondary patterning behaviour. In brief, the Gierer–Meinhardt dynamics are not nearly as good as the Brusselator for spacing repeated parts regularly and, somewhat surprisingly, continuing to 'know' the linear wavelength when behaving non-linearly. But the Gierer–Meinhardt dynamics are very good at telling repeats of an activation event to keep away to a much greater distance than the range of activation. This is very promising for explaining single-organ location in animals.

But while reaction–diffusion patterns in 1-D show significant departures from their starting forms as regular repeats of sine waves, the variability of patterning becomes much greater in 2-D systems much larger than the pattern wavelength, and therefore having a large number of pattern repeats. First, there is the matter of whether 2-D patterns will be stripes or spots, both of which are common in animal coat patterns. I take this up in Section 9.2. Then, for the spotted cases, computations show that even a mechanism with complete activator–inhibitor–diffusion Turing properties will sometimes generate patterns that are only partly ordered, like those used in Figs. 6.1 and 6.2 to illustrate the Wigglesworth mechanism.

Here, the diffusivity ratio $n = D_y/D_x$ is significant; the higher it is, the more closely the system behaves to one with immobile activator spots, and the more irregular the patterning looks; but also, Gierer–Meinhardt patterns become disorderly at lower values of n than Brusselator patterns (Fig. 9 of Lacalli 1981, repeated as Fig. 9.5 of Harrison 1993; Figs. 2 and 4 of Holloway and Harrison 1995). In the following subsection, I describe the work reported by Sick *et al.* (2006) on mouse hair follicle patterns, which show similar moderate disorder both experimentally and in computations on a model which is the closest yet reported to definitive identification of the two morphogens in a Turing mechanism.[5]

6.3.5 Activator and inhibitor distributions: what proves what?

Whether or not this brief listing of components of a Turing mechanism and pictorial statement of their dynamics in Figs. 6.7–6.9 is sufficient to give anyone the slightest inkling of how they work to form pattern, I hope it has clarified a danger in trying to put this kind of theory to experimental test. Reaction–diffusion pattern formation is bristling with counter-intuitive properties. For instance, I have heard a very competent developmental biologist say: 'I've found an activator at one end of the system; now I'm going to look for an inhibitor at the other end.' But a glance at Fig. 6.5B, with the solid and broken lines interpreted as X and Y concentrations, shows that the maxima of those two substances coincide. One should expect to be able to harvest the greatest amount of inhibitor exactly where you have just found the most activator. Indeed, this is obvious from Wigglesworth's concept, in the outward-diffusing inhibitor form. It is the activator spot that generates

---

[5] See also Digiuni *et al.* (2008) for Turing morphogens in plant development.

the inhibitor. This gets worse. I have distinguished, in Section 6.1, between inhibition and depletion models. They work in essentially the same way, dynamically. But (and again this is obvious in the Wigglesworth mechanism) the depleted substance is indeed lowest at the activator maximum and highest at the activator minimum. To know whether the second substance in a waveforming reaction–diffusion mechanism is going to be in phase or 180° out of phase with the first substance, one has to know whether the mechanism is activation–inhibition or activation–depletion; or one has to be ready to settle that by experimental measurement. Meinhardt (1984) put the dangers thus:

> Since at least two substances are required which act in an antagonistic and non-linear way with each other to form a pattern, these systems have counter-intuitive properties. For instance, the removal of an inhibitor from a tissue can lead to an increase of the inhibitor concentration (due to the triggering of a new activator maximum). Or, in an attempt to isolate the activator, isolated activated cells would appear as non-activated since the inhibitor can then no longer escape into non-activated surroundings. The accumulating inhibitor would rapidly switch off the activator production. Without knowledge of the underlying principles on which pattern formation is based, the chance would be high that a successful experiment would be incorrectly interpreted as a failure. We hope that our models can help to design appropriate experiments.

Let us consider a specific example: chicken feather buds form as a regular hexagonal array of spots. This seems to indicate a pattern-forming mechanism with stronger geometrical ordering properties than a Wigglesworth inhibitory field. In normal development, the pattern is formed one row at a time. It has generally been believed that this is necessary for the achievement of the good order. (Provoked by quite a different phenomenon, in *Acetabularia* whorls, I have suggested (Harrison 1982) that Nature commonly gets control of pattern by reducing spatial dimensionality.) But studies by Ting-Xin Jiang *et al.* (1999; Han-Sung Jung *et al.* 1998) have shown: first, that dissociated mesenchymal cells, reaggregated and reconstituted with non-labelled epithelium, reproduce the whole hexagonal pattern simultaneously, not row by row; second, that a number of gene types for which the protein products are known to be important signalling molecules in diverse organisms are involved as activators and inhibitors in this patterning, particularly: sonic hedgehog (Shh), bone morphogenetic proteins (BMPs) and fibroblast growth factors (Fgfs). From the 1998 paper: 'The fact that both the proposed activators (Shh, Fgf4) and inhibitors (BMP2, BMP4) are colocalized in the feather primordial regions, rather than having the activators in the primordial and the inhibitors in the interprimordia regions, favors a reaction–diffusion mechanism as proposed by Turing.' This, I think, is a valid conclusion, given that the BMPs appear to be true inhibitors, not inhibitors de facto by depletion, and that the pattern is more strongly ordered than those that attracted Wigglesworth's attention. To my mind, the study of this event is getting quite close to definite identification of a Turing morphogen pair.

Such substances are being mentioned also (sometimes together with transforming growth factor Tgfβ and things in the 'Wnt signalling pathway') as possibly involved in dynamic patterning mechanisms in gastrulation (Section 9.4), limb morphogenesis (Section 9.5.1) and branching lung morphogenesis (9.5.2). The possibility that mammalian hair follicles might be patterned by a Turing mechanism was proposed by Nagorcka and Mooney (1982, 1985).

Sick *et al.* (2006) have quite forthrightly and prominently announced that 'WNT and DKK determine hair follicle spacing through a reaction–diffusion mechanism', publicized and assessed, chiefly supportively, by Maini *et al.* (2006) as 'The Turing model comes of molecular age'. Following these accounts, having been published in a journal of high standards and strongly connected to experimental work on mice with computations on reaction–diffusion, it can no longer be stated that no Turing morphogen pair has ever been identified. (Reminder to readers unfamiliar with the arcane conventions of writings on genetics: all capitals, WNT, DKK, proteins; *Dkk1*, the gene coding for the protein.) Here, it behoves me to respond to my question 'what proves what?' by acting for a paragraph or so as devil's advocate and looking for whether there are any weaknesses in the evidence for this definitive identification of a reaction–diffusion mechanism.

The definite strengths are: WNT is definitely an activator for induction of hair follicles; 'Expression of the WNT inhibitor *Dkk1* is directly controlled by secreted WNTs'; and WNT proteins are substantially larger than DKKs, giving the expectation that the diffusivity of the activator will be definitely less than that of the inhibitor. These things alone almost demand that reaction–diffusion mechanisms should be introduced into any serious discussion of the hair follicle patterning. Sick *et al.* (2006) go on to derive from the reaction–diffusion hypothesis testable predictions regarding 'the outcome of experimental alterations of activating and inhibitory functions'. To understand fully what they are doing here, it is necessary to read their supporting online material that gives the reaction–diffusion equations they use in their model of the mechanism. For these, they choose the Gierer–Meinhardt form for the dependence of rates of formation of activator and inhibitor upon the concentrations a and h of those substances:

$$F(a, h) = a^2 / \left[(K_h + h)(1 + ka^2)\right], \tag{6.6}$$

where $F(a, h)$ is the term dependent on concentrations of activator and inhibitor in the rates of formation of those two substances. $(1 + ka^2)$ is a 'Hill term' representing saturation of the rate of production of activator at high activator concentration. The $F(a, h)$ expression is a good choice for the ways in which rates may depend upon a and h, but it is by no means the only possibility. The Brusselator (Section 5.2.2) would give quite different functional dependences, and there are many other possibilities. But the authors indicate that their reaction–diffusion equations reflect characteristic features of the WNT signalling pathway, particularly the non-competitive inhibition of WNT signalling by DKK.

They predict from their model that moderate overexpression of the activator (WNT) should increase follicular density, i.e. decrease interfollicular spacing, while moderate overexpression of inhibitor (DKK) should have the opposite effect; and they find some experimental data in agreement with prediction, especially in regard to inhibitor expression. Such dependences, however, are not necessarily the same for every form of $F(a, h)$ that could be written in place of Eq. (6.6) in a reaction–diffusion model, and dynamics other than reaction–diffusion (Maini *et al.* 2006 mention chemotaxis) could have some similar patterning properties. These authors write that 'Now that WNT and DKK have been identified as possible morphogens … the key requirement … is that the results of such experiments are used to test and refine models.' They mention tests involving measuring parameters such as rates of production, decay and diffusion coefficients that would determine whether the system actually is of Turing type. They feel that such things are necessary so that 'this would then be the first definitive example of the Turing model in biology'. An important implication of this discussion is that, even when Turing dynamics have been suspected with high probability in a system, a lot of work both experimental and theoretical is still needed to tie down the identity of the mechanism with full assurance. There is no question of doing the full job by a quick scan of the kinds of interactions that seem to be present in a large network.

Here, of course, a Turing model is being thought of in the narrow sense, as strictly involving two molecular species. In the present book, I have indicated from the preface onwards my view that there is a broader view in which Turing mechanisms can be considered as the presence of particular kinds of dynamics, perhaps on the molecular scale and perhaps on larger scales. The reader may find it useful to contrast this example with the report of Kondo and Asai (1995) on observed changes in the macroscopic features of skin patterns on an angelfish, which received the cover headline 'Turing patterns come to life' (Section 9.2). The two examples taken together give, I think, a useful perspective on the matter of 'what proves what?', or what is to be regarded as evidence for what.

6.4 IS REACTION–DIFFUSION ONLY TURING?

6.4.1 Only one morphogen: the exponential gradient

Science is replete with very general concepts and with concepts of very restricted application. Which is reaction–diffusion? To me, it is obviously a very general term; it signifies anything to do with rates of reaction and rates of random movement of molecules. I am often worried that the term may have become identified in many people's minds with the properties of only one pair of differential equations: Turing's equations.

And further, it is possible to become confused with the terminology and begin asking: could this behaviour be Turing, or Gierer–Meinhardt, or Brusselator, or …? But in fact, Turing is a fairly general category that includes both of the other two just named, and reaction–diffusion is a much more general term than Turing.

For example, it is quite common in biology to postulate the existence of an exponential gradient of the concentration C of some substance,

$$C = C_0 \exp(-ks), \tag{6.7}$$

where s is the distance along a 1-D system. In my experience, the probable existence of such gradients seems to be accepted by most biologists; but the origin of such a gradient is much less commonly mentioned.[6] It is in fact well known and quite simple, and it is definitely within the category of reaction–diffusion: a substance is produced at a point source with concentration regulated to be always C_0. It diffuses from the source, at diffusivity D and decays with first-order rate constant k_d. (Such decay is a common property of most proteins, regardless of whether there is a specific proteolytic enzyme present.) Then:

---

[6] Though see Lander (2007) and Bergmann et al. (2007) for recent discussions.

rate of change of C = decay reaction rate + diffusion rate

$$\partial C/\partial t = -k_d C + D\partial^2 C/\partial s^2. \tag{6.8}$$

If we now ask what steady-state the gradient will settle down into by writing $\partial C/\partial t = 0$, then the differential equation to be solved is:

$$C = (D/k_d)\partial^2 C/\partial s^2. \tag{6.9}$$

The appropriate solution to this equation is the gradient given by Eq. (6.7), if

$$k = (k_d/D)^{1/2}. \tag{6.10}$$

The best-known experimentally observed instance of such a gradient is that of the Bicoid protein in the *Drosophila* embryo, which is discussed in Chapter 8. Another is the gradient of retinoic acid specifying the order of digits towards the end of limb morphogenesis (Section 9.5.1).

6.4.2 More than two morphogens

Reaction–diffusion dynamics that have been successfully used to account for the generation of various kinds of pattern are not restricted to only two protagonist substances. Lacalli (1990) used a four-morphogen model devised from two side-by-side models of Brusselator type to discuss the striped patterns of 'pair-rule' genes that mark the onset of segmentation in the fruit fly *Drosophila*. (See Chapter 8 for discussion of the controversy over whether these patterns have anything to do with reaction–diffusion at all.) In his book on sea-shell pigmentation patterns, Meinhardt (1995, Fig. 7.2 and Eq. (7.1a–f)) used reaction–diffusion equations for six diffusible components to model the formation of the pigmentation pattern on the cone shell *Conus marmoreus*. When I pointed this out in passing to a theoretical chemist, he remarked 'That isn't very many, for a pattern as complicated as that.' Holloway and Harrison (1999a, Fig. 9) used an idea from that set of equations in our mechanism for control of successive dichotomous branchings in the algal genus *Micrasterias*.

In the sea-shell pigmentation phenomenon, Meinhardt has seized upon a living system that does part of the scientist's work for him: it plots a 2-D space–time graph of a 1-D patterning process that most often never reaches a steady-state. Meinhardt (1982) has quite a liking for this kind of representation of computational results. Many of the gastropod molluscs do it by forming the shell, and its pigmentation, one row at a time from a narrow edge of an organ called the mantle. Thus each row is a record of the pattern along the mantle edge at one time, and successive rows as they are added build up the time axis. (Some gastropods, however, make the whole 2-D pattern simultaneously.)

Modelling as complex as the six-equation *C. marmoreus* instance requires that people with some expertise in mathematical or computational work be involved in the project. But some properties of straightforward two-morphogen models should be usable by experimental biologists in quite a general way. The chapter entitled 'Pictorial reasoning in kinetic theory of pattern and form' in Harrison (1993) was quite largely devoted to unequal branchings of cells in filamentous cyanobacteria (*Anabaena, Chaetomorpha*) and plant roots (the water plant *Azolla* (Gunning 1981, 1982; Barlow 1984)), showing chiefly how changes in boundary conditions for a reaction–diffusion system brought about by one new division plane could determine the location and orientation of the next one. The thinking needed for these examples could be related entirely to picture-drawing, without use of equations, provided only that one accepted the simplest generalities of how a sine wave pattern should respond to specification of a new boundary as a trough or a peak or a node in the pattern.

In Chapter 1 of Harrison (1993), I discussed the philosophy of turning abstract mathematics into 'thinking in concrete terms' with the following analogy:

> The Schrödinger equation ... is quite necessary to show, for instance, that an O_2 molecule is held together with the strength of a double bond, and yet contains two unpaired electrons. Many chemists of the molecule-making majority, however, do not spend their time maintaining fluency in the language of the Schrödinger equation, with its full panoply of spherical harmonics, Legendre polynomials, orthogonal sets, Hermitian operators, and so on. Rather, they would "prove" the foregoing statement about O_2 pictorially, by drawing an energy-level diagram and putting electrons into the levels. Such diagrams are powerful, permitting one to conduct many correct derivations without having to resort to equations.
>
> But in the end, the equations have led to the diagrams and are the only justification for them.

The question I was trying to raise was: are we now at a stage where developmental biologists can similarly use the properties of dynamically generated patterns without having to dig into the equations? Clearly there is one big difference between this biological case and the quantum chemistry analogue: the latter is based on widely accepted theory, while the former still awaits anything remotely like the same level of acceptance. Here there is a strategic problem: quantum mechanics of atoms could be established firmly from physical phenomena, especially spectroscopy, and then brought securely into chemistry. Dynamical formation of pattern in developmental biology can be tested only by people sufficiently convinced of its promise that they are ready to use it in their thinking about experimental data as they accumulate them. That is the reason for my continued evangelism on this topic.

7

Classifying developmental theories as physical chemistry

7.1 STRUCTURE, EQUILIBRIUM, KINETICS

Every species of scientist seems to have particular preoccupations displayed by all members of the species and quite unknown to anybody else. For physical chemists, one of these is: 'Are we talking about kinetics or equilibrium?' Part of the difficulty here is that the word 'equilibrium' is used in a number of different senses by different kinds of scientist. Physical chemists use it in a rather strict sense, to mean 'thermodynamic equilibrium'. This is a state of a system in which no macroscopic observables are changing with time, and there are no flows of energy or material through the system. In terms of biological systems, this means definitely dead; and such terms as 'equilibrium evolution' therefore tend to make physical chemists shudder.

Chemical reactions in a closed system, i.e. one to which no changes are being made by adding or removing material across its boundaries, proceed until equilibrium is reached. At that point nothing further happens. If, however, reactants are continuously being added and products removed, it is possible for the contents of the reaction vessel to reach time-independent concentrations and spatial distributions, the constancies of which are entirely dependent on continuous supplies and removals. These are known as 'out-of-equilibrium steady-states', and any pattern that one believes to be in the dynamically generated and dynamically maintained category is thereby counted among them. When we talk about it, we are discussing kinetics, not equilibrium.

With this clarification, I can now propose that all physical chemistry, like Caesar's Gaul, is divided into three parts, 'structure, equilibrium, kinetics', three words I used as the title of Chapter 4 in Harrison (1993). And all of these may play a part in developmental patterning. Many instances of patterning belong predominantly to one of these three categories; and for these, in my 1993 book (Chapter 8), I devised a quasi-Linnaean classification of developmental theories in which these three words defined the phyla, with, for instance, four classes in the kinetic phylum named: reaction–diffusion, mechanochemical, self-electrophoretic and 'complex intercellular signals, e.g. assembly of the central nervous system'. I think I may have intended this classification more to provoke than to inform, because its framework was rather unlike the scaffolding of the thinking of most developmental biologists. And there are flaws in the analogy to taxonomic classification: first, while a given developmental event will most usually belong primarily to one of the three main categories, they are not totally mutually exclusive. In the earlier book, I included several problems asking the reader to discuss the structural, equilibrium and kinetic aspects of a number of events. My objective is not that a 'tree of mechanisms' should be more and more precisely constructed, but that thinking about developmental events in the manner of physical chemists should become an expanding part of developmental biology, and not only in the minds of physical chemists. Second, I have already tried to establish that my obsession with reaction–diffusion equations, largely of the two-morphogen Turing type, arises because I think that they embody a dynamic principle much broader than just two-substance molecular dynamics, which indeed can slop over into other 'classes' of the kinetic 'phylum'. In Section 9.2 I mention that my student, Michael Lyons, started tackling the stripes-versus-spots question for reaction–diffusion, but ended up finding the very same stability conditions applicable to an instance of 'complex intercellular signals' in the eye–brain connection.

The terms 'self-assembly' and 'self-organization' are worthy of a bit of clarification. As I understand the usage of the former, nearly everybody uses it as identical to my 'structure' category, i.e. the building-block or jigsaw-puzzle formation of a large object by the geometrical fitting together of small pieces. 'Self-organization' can, I think, be used in more diverse senses. I use it to mean what I have addressed in Chapter 1: any mechanism for pattern formation in which the Organizer is integral with the thing being organized.

Most of this book is concerned with the continuing (and to my mind still far from sufficiently pursued) promise of kinetic explanations. Structure is not my topic. But, though I have above identified equilibrium with the definitely dead, matters related to the approach to thermodynamic equilibrium sometimes turn up in the darnedest places in living systems that are not about to die. I devote the remainder of this chapter to a couple of examples of these. While presented primarily in the context of equilibrium, these could be regarded as two more problems in my category of 'discuss the structural, kinetic and equilibrium aspects'. Even more generally, the discussion illustrates my principal thesis in writing this book and its 1993 predecessor: that the mode of thinking of classical physical chemists, as it developed from, say, the 1850s to the 1950s in regard to events on spatial scales above the molecular, should become a much larger component than it now is of the future of developmental biology.

7.2 JUST FOR A CHANGE, EQUILIBRIUM: OR IS IT?

7.2.1 A phase transition with a beating heart

Scientific literature in any major field of work is today so vast that one is doing well to get references to all the significant workers into any one account, never mind all the significant references to one person or group. But for the latter, which is most important: to give priority by quoting the earliest paper, to give some intermediately dated reference that contains the most significant material, or to give the most recent reference, which may be adequate for tracing backwards? I have noticed that the work of Malcolm Steinberg on cell sorting by differential adhesion is often given less than its due by being referred to only in regard to papers in the 1960s, when the most significant papers started with Phillips and Steinberg (1969) and most particularly Steinberg (1970), and continued to the summary in Foty *et al.* (1996), from which I have extracted Fig. 7.1, and the title of which 'Surface tensions of embryonic tissues predict their mutual envelopment behaviour' essentially gives the whole message. This is that, when one tissue places itself inside another, ideally as a packed sphere of cells inside a spherical shell of the other tissue, this geometrical arrangement, including the matter of which one is the tissue that goes inside, is decided by relative strengths of attractive forces between cells of each type. This is the 'differential adhesion hypothesis'. In such a case, we are seeing the final arrangement as one of minimum energy, and therefore of thermodynamic equilibrium. (For such small numbers as those of observable cells, any entropy of mixing factors are negligibly small, and relative energies can be identified with relative free energies.) The topic is equilibrium in a system of two fluid phases, with what I call the 'cell-as-molecule' approach. Eventually, following the identification of various classes of cell adhesion molecules, Foty and Steinberg (2005) used substances from one of these classes, the cadherins (calcium-dependent

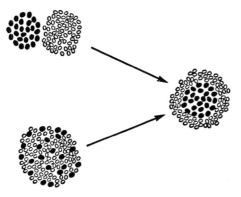

Figure 7.1 Sketch of two types of experiments on pairs of cell types of embryonic chick tissues, as used in the work of Steinberg (1970) on cell sorting. Upper arrow: an engulfment (or aggregate fusion) experiment. Pieces of two different tissues in culture are threaded together on one skewer. One piece moves to engulf the other. Lower arrow: a cell-sorting experiment. The two tissues have been disassembled (by proteolytic enzymes) into separated cells, which have been randomly mixed into a single aggregate. They move to form a central sphere and an enveloping spherical shell. For the same two tissues both experiments almost always gave the same result, indicating that the result is most probably in equilibrium.

adhesion molecules) to show that the surface tensions of cell aggregates transfected with these substances were a direct, linear function of cadherin expression level. But they continued to stress that their conclusions on cell sorting, tissue spreading and segregation of cell types relate to 'the physics governing these morphogenetic phenomena . . . independently of issues such as the specifics of intercellular adhesives'.

Two preliminaries before I get to the Steinberg work: first, when a liquid system, usually a solution of two substances, separates into two phases, what we usually see is two layers, one above the other, with a horizontal flat interface between them. The lower layer is the denser one, i.e. this arrangement is determined by gravity. But if a phase separation were carried out in zero gravity, what we would expect is that instead of two layers, the final arrangement would be a sphere of the substance with the more strongly attractive molecules inside the other. (This is an approximate description; usually neither phase is one pure substance. Each has both components, but in very different proportions.) A zero-gravity experiment on such a system was in fact carried out during a space shuttle flight. It was designed by Donald E. Brooks and performed by Senator Jake Garn from Utah. The result was the expected one: for fluids in a thin square cell, the phase separation happened just as well as it does on Earth, but one of the phases formed a circular island within the other.

Second, the phenomenon of cell sorting first attracted attention in relation to the three layers, ectoderm, mesoderm and endoderm, the formation and proper arrangement of which are crucial to most early animal embryonic development. Tissues of amphibian embryos, in culture, can easily be made to fall apart into separated cells, by making the culture medium alkaline. When random mixtures of cells from ectoderm and mesoderm were made, and the pH restored to normal, cells would move to form separate aggregates of the two types, with the mesodermal aggregate enveloped by the ectodermal. Townes and Holtfreter (1955) reported this. If cells from all three germ layers are used, there is envelopment with ectoderm as outermost and endoderm as innermost. Holtfreter interpreted this in terms of 'selective affinities' of cells for others of the same or different types.

Steinberg (1970) made this concept more precise as 'differential adhesion' analogous to the differential attractions of molecules in a liquid solution of two or more molecular species. He did experiments using six types of tissue from chick embryos. These are more difficult to use than the amphibian embryos, because it is necessary to use proteases to separate the cells. I shall describe the statistical treatment of the 1970 data with six tissue or cell types, A, B, C, D, E and F. (The meaning of my title for this section is that one of the cell types was heart muscle cells, and when they formed an aggregate it started autonomously beating, a well-known characteristic of such cells, but not quite expected by a physicist studying phase transitions in fluids.) These six types were used in all 15 possible pair combinations in two types of experiment (Fig. 7.1): engulfment, in which pieces of the two tissues were put in contact by being threaded together on a skewer; and cell sorting, using random mixtures of separated cells. For the same two tissue types, both types of experiment always gave the same result. As Steinberg pointed out, when the same result is obtained by two different paths, one has a strong indication, though not a proof, that the result is determined by equilibrium. Next, Steinberg was looking for whether or not the results had a transitive property, establishing the existence of a hierarchy. The transitive property is that, if $A > B$ and $B > C$, then $A > C$. Here, the $>$ sign can be any difference in behaviour of the two items when paired. In this case, $>$ can be 'goes inside'. One has to be careful about statistical significance. For only three tissues, A, B and C, there are eight possible sets of results of pair-competition

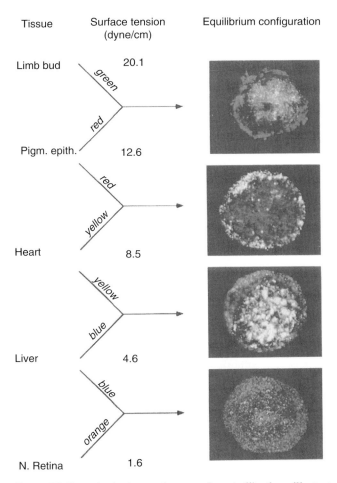

Figure 7.2 For pairwise interaction experiments, like those illustrated in Fig. 7.1, between two embryonic chick tissues from a set of five. The results for five of the possible ten pairings are shown in the right-hand column. The tissues are listed in order of their surface tensions, which is also their hierarchical order in the envelopment phenomenon. The five pairings shown are for adjacent tissues in that order. From Foty *et al.* (1996) with permission.

experiments, and six of them correspond to hierarchies. There is a 75% chance of 'proving' the existence of a hierarchy when the results are in fact random, with no difference in the property driving them. For six items, this chance is reduced to 2.2%. To that level, Steinberg's results established the existence of a hierarchy for six tissue types.

Fig. 7.2 shows results of a later experiment with only five types of tissue, from Foty *et al.* (1996).

What such a result establishes is the existence of a property of each tissue (or cell) type that should be quantifiable; the one that goes inside in the final configuration is the one with the higher value of that property. A common chemical example is the experiment of dipping a metal into a solution of a salt of a different metal, and seeing that the first metal dissolves and throws the second out of solution. This gives us a replacement series as the hierarchy: Zn replaces Cu; Cu replaces Ag; and Zn replaces Ag. This gives Zn > Cu > Ag. The quantifiable property is redox potential (oxidation potential for the order given, reduction potential if one set up the inequality in reverse, which is just a matter of what one takes > to mean.) For the two-phase separation of liquids, the property is either surface tension of each pure liquid, or energy of cohesion between molecules of the same type. For an aggregate of cells of any one type, Steinberg was able to measure an analogous property by measuring quantitatively how much an aggregate stayed rounded or flattened in a centrifugation experiment (Phillips and Steinberg 1969; Foty and Steinberg 2005). The order of surface tensions indeed corresponded to the order of the hierarchy of cell types. For artificially introduced cadherins, Foty and Steinberg can now correlate the surface tension with the amount of the cadherin expressed in the cells. (These are not the only known type of cell adhesion molecules, and the experiments do not prove that they are the kind active in embryonic development.)

In the cell sorting and engulfment experiments, two types of movement are seen: random, diffusion-like movements of individual cells (usually called 'passive movement' in biology) and coordinated, orderly flows of streams of cells of one type. Both happen in animal embryological events; here, see particularly the account of gastrulation in Section 9.4. But now I have moved from the equilibrium end-results to the kinetics by which they are achieved, and hence quickly jumped one phylum in my classification of developmental mechanisms. Attempts at computational modelling of cell movements have been made since the 1970s. For the phenomenon of cell sorting, two categories should be distinguished: (1) modelling that seeks to find the minimum free energy configuration, regardless of what movements of cells need to take place to get to that point, e.g. Goel and Rogers (1978); (2) modelling that seeks to display the kinetics of cell movement. For an early example of this, see Matela and Fletterick (1980), described in Chapter 4 of Harrison (1993). (I am here using 'early' to mean a quarter-century or more ago; this seems appropriate to the topic, but goes against the grain – I prefer to use 'early' to mean at the very least a century ago.)

The 'early' attempts were a good start, but had a number of defects, of which the two most important were: (1) the computations tended to get stuck at configurations that were low-energy, but not lowest energy, i.e. several small patches of the highly cohesive species instead of one big patch inside the continuum of low cohesion; (2) a thing endowed with the ability for complex movement, the biological cell, was represented by one indivisible object. Recent (i.e. post-1990) greatly enhanced success in modelling cell sorting owes its success principally to two things: (1) it can be tied to a much longer tradition in physics, the queen of sciences (and is therefore royally endowed); (2) it models a biological cell with an average of about eight pieces, in changeable geometrical arrangement and slightly changeable but close to constant number and therefore cell volume. This is a minimal representation of the complexity and flexibility of a cell, but a lot better than just one piece. It is enough to do the job of not permitting the configuration to stop at several small aggregates of cells of the more cohesive component, but letting it see that one large aggregate is better.

The successful cell sorting model is that of Graner and Glazier (1992); Glazier and Graner (1993). This kind of modelling is clearly described in the PhD thesis of A. F. M. (Stan) Marée (2000), who is now using it in modelling the kinetics of vertebrate gastrulation (Section 9.4). The modelling game (in 2-D; it can be extended to 3-D) is played out on a lattice of square sites (Fig. 7.3), upon which a biological cell will occupy several (about eight) sites. It is derived from the Ising model, from 1925, very well known in physics, for magnetic properties arising from interactions of 'up' and 'down' electron spins. Biologists may sometimes be discomforted by hearing physicists, when discussing such modelling for biological events, almost unconsciously slipping into talking about 'spins', a word irrelevant to the matter at hand, but definitely whence their models derive. Potts (1952) extended the Ising model to more than two possible states per site, and the biological model may be given his name or, more usually, those of Glazier and Graner.

Fig. 7.3 shows a portion of a lattice, in which unshaded sites are occupied by culture medium M (or extracellular medium, its natural equivalent). Two levels of shading are used to distinguish two types of biological cell, t and p. Across one line (side of square) between two sites, there are five possible juxtapositions: t,M; p,M; t,t; p,p; t,p. Each of these corresponds to a value of free energy per unit area of interface. These can be summed for a total excess free energy of the whole system due to the interfaces it contains. In a step of a computation, a site is chosen at

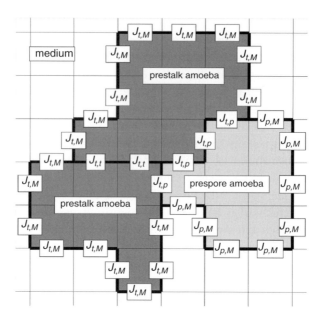

Figure 7.3 The model of Graner and Glazier (1992) for cell sorting, as depicted by Marée (2000). On the square lattice, a biological cell occupies several squares, and the occupied ones can be moved individually to represent amoeboid movement (see text).

random, and the state (t, p or M) of one of its neighbours is copied into it if this gives the system lower energy, or with a probability $\exp(-\Delta H / kT)$ if the energy change ΔH is positive. Clearly, this kind of copying will sometimes change the volume of a cell. This is not likely to happen for real biological cells, and reality is restored in this respect by putting an energy penalty into the total energy formula, so that energy is increased when cell volume changes. This algorithm gives a good account of cells moving in an amoeboid manner to minimize the energy of the system, and it allows the destruction of small aggregates that give low energy but not the true minimum, in favour of one large aggregate that is the minimum.

What is the balance between structure, equilibrium and kinetics in theories of cell sorting? The final result is a structure in the sense that cells of the same kind are grouped together; but the grouping is like that of molecules in a fluid, not a fitting together based on shapes of the individual cells. The system ends in thermodynamic equilibrium in respect of the energetics of cell-to-cell contact, though each cell is internally a living thing, and therefore out of equilibrium. But sorting is a process, involving movement and

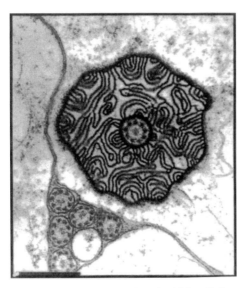

Figure 7.4 A pattern, produced within a living organism, and on a much smaller spatial scale than most other patterns discussed in this book. Does the picture contain evidence for a dynamic mechanism of patterning (see text)?

hence kinetics in any complete explanation of how it happens, as in the Graner–Glazier method for computations.

Hence, Steinberg can look from equilibrium towards structure in calling cell sorting 'self-assembly'; and Marée can look from equilibrium towards events that are more definitively kinetic in taking over the Graner–Glazier method into instances of cell streaming (Section 9.4, gastrulation and slime mould culmination).

7.2.2 A phase transition in a condensing nucleus?

First, look at Fig. 7.4 in uncritical admiration of the pattern without thinking too much about what it might be. This one, and many of its relatives, were brought to me by a zoologist, Harold Kasinsky, with the question: could there be a dynamic explanation of how these patterns are formed? When I got interested in this, my right-thinking friends (i.e. believers in reaction–diffusion) said: 'You're not going to try to explain *that* by reaction–diffusion, are you?' The snag is the spatial scale. The pattern repeat spacing between, e.g. the middles of two adjacent black stripes, averages about 40 nm, i.e. something about 250 times smaller than the repeat spacing of about 10 μm that is the norm for developmental pattern formation. I could almost claim to be doing some kind of nano-science, considering this pattern. It's at about the top of the nano-range. For reaction–diffusion, I have indicated that time and space scales are linked; the smaller the spacing, the shorter the time of pattern formation (Section 2.3). For this tiny value, one might expect a time of pattern formation of a fraction of a second. But this pattern arises in a process that takes, *overall*, something like a couple of weeks. 'Overall' implies that we have no way of finding out what fraction of this period is occupied by the formation of this particular pattern; but a fraction of a second seems unlikely. This pattern, like anything else biological on this spatial scale, cannot be observed in vivo as it forms. It is seen only in transmission electron micrographs (TEMs) of dead material.

The pattern has, however, a striking feature that strongly suggests the involvement of diffusion in its generation. The stripes all run into the boundaries (both inner and outer) of the patterned region perpendicularly, like the stripes running to the long edges of the rectangular vessel in Fig. 6.5. In that context, I have pointed out that, where the boundaries are 'no-flux' for the diffusing species, the concentration gradient at the boundaries must be zero, which implies that stripes must run either perpendicular or parallel to the boundaries, where they meet those boundaries. (In the set of patterns from which Fig. 7.4 is taken, we also have examples of the parallel orientation.) Of course, there could be a structural reason for the orientation in the supermolecular or nano-scale architecture of the stripe–boundary junction. But in the absence of any knowledge of such architecture, it is reasonable to take this picture as giving evidence for a mechanism including diffusion.

Fig. 7.4 is in fact a TEM of a cross-section of a nucleus of a spermatid of a whelk (gastropod mollusc), *Murex brandaris*, a denizen of the Mediterranean Sea. The sample was prepared in the usual way by glutaraldehyde fixation and heavy metal staining. The dark stripes are chromatin (DNA interacting with sperm nuclear basic proteins (SNBPs)) and the intervening light regions are nucleoplasm, probably having diverse constituents from proteins to small ions. The outer and inner boundaries are bilayer membranes. The word spermatid designates a cell after the reducing divisions; i.e. haploid, with one copy of each chromosome, and not yet a mature sperm. Since sperm are produced by both plants and animals, ranging from mosses and protozoa to humans, it is hardly surprising that there is great variety in the details of spermiogenesis. It commonly involves a shrinkage of the cell nucleus to between 1/20 and 1/200 of its initial volume; in the final state the chromatin occupies most of the remaining volume, i.e. it is condensed, and the nucleus looks completely dark in TEMs. The formation, at intermediate stages on the way to condensation, of patterns as spectacular as Fig. 7.4 is not general. The examples I was shown for the patterning question were a few internally fertilizing molluscs and insects (Harrison *et al.* 2005),[1] both from work of the co-authors of that paper and from earlier literature (Chevaillier 1970). My first instinct was to lay a ruler on the pictures that were spread out before me and find a constant-spacing phenomenon for the distance from the middle of one dark stripe to the middle of the next across the nucleoplasm, a supposedly fluid region with no known structure. The spacing stayed constant through substantial changes in the appearance of the pattern.

I have already given a reason why this cannot be reaction–diffusion patterning, in terms of space versus time. Another good reason is that we were not looking at patterns of reaction intermediates in the middle of continuous processing of reactants into products. The DNA isn't doing that. In fact, the whole nucleus is effectively shutting down its chemistry and assuming the most compact form for the journey that a tiny minority of the sperm will successfully complete to insertion of the nucleus into the egg. The condensed sperm nucleus is perhaps analogous to a well-folded parachute, ready for instant action when unfolding becomes necessary. Also, it may be the closest that anything in a living organism gets to being at thermodynamic equilibrium without actually being dead.

All these considerations led me to think about a mechanism for transient patterning of two fluid phases on the way to separation, known as 'spinodal decomposition', the theory of which was devised by Cahn (1965) following a treatment by Cahn and Hilliard (1958) of the free energy of a system with a concentration gradient in it. Hence the dynamic equations for this kind of patterning are often called the Cahn–Hilliard theory. (Further terminology: I understand the word 'spinodal' came, among scientists in the Netherlands following van der Waals, in some light-hearted manner from the name of the philosopher Spinoza.) A solution of two substances (X and Y) may be quite stable as a single phase at one temperature; but if the temperature is lowered, it may become unstable relative to a system of two phases, one largely X with a little dissolved Y, the other largely Y with a little

---

[1] See also Martens *et al.* (2009).

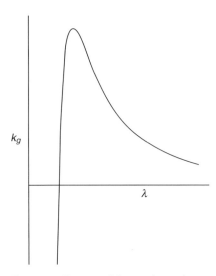

Figure 7.5 Exponential growth rate k_g versus wavelength λ for a sine wave pattern of concentration during the phase separation mechanism called 'spinodal decomposition', from Cahn (1965) with permission. Compare the curves for Turing mechanisms in Figs. 2.6 and 6.7, also Fig. 7.7 of Harrison (1993), which illustrates that a Turing k_g does not always go negative at long wavelengths.

dissolved X. The question arises of how the system will move to achieve this new equilibrium state: does it need nucleation of small droplets at the two new equilibrium compositions, or can the uniform concentration gradually acquire increasing concentration gradients? Both paths are in fact possible, dependent upon the overall composition of the system and how free energy changes with that composition. The details of the former path, nucleation followed by coalescence of small droplets, was, as mentioned in Section 7.2.1, one of the problems in computational modelling of cell sorting as a phase separation phenomenon. For the latter path, gradual increases in inhomogeneity throughout the system, Cahn showed that the earliest stage would be a waveform with a similar growth-rate versus wavelength variation to the 'band-pass' property of Turing waves; compare Fig. 7.5 with Fig. 6.7.

The big contrast between Cahn and Turing patterning lies, however, in what determines the wavelength of the fastest-growing pattern, and what this says about spacing versus time. Turing patterning depends on a saw-off between diffusion and reaction; to get a linear dimension for a wavelength, the two quite different time dependences of these have to cancel out (expression with Ds and ks in Fig. 6.7). Cahn patterning has only diffusion as a time-dependent quantity. His theory leads to an equation for growth rate k_g:

$$k_g = -M(\partial^2 f/\partial c^2)\omega^2 - 2Mk\omega^4 \qquad (7.1)$$

where M is diffusive mobility in the fluid, f is free energy per unit volume, c is concentration of one of the components, k has to do with how a concentration gradient changes the free energy, and ω is, as in Chapter 6, 2π / wavelength. The point here is that when one does a bit of high school calculus on this equation to find the wavelength for maximum k_g, one of the M-dependent terms cancels out against the other, and one therefore gets the same wavelength regardless of how fast or slow the diffusion is. This theory can cope with a pattern with tiny spacing that forms very slowly (or any other combination of size and speed).

And again, I must inflict my own quiz upon myself: structural, equilibrium and kinetic aspects? Well, the Cahn theory is for approach to a phase equilibrium, but doesn't follow the event all the way. It describes, kinetically, the occurrence of a transient intermediate stage; transient, because the pattern is not kinetically sustained by inputs to and outputs from the system, as is necessary for Turing pattern. And structure? Rather ghostly, as concentration gradients in the system; the pattern wavelength does depend on the size of the molecules in the system, but as modified by the range of their energies of interaction, not as a hard-contact size and shape.

7.3 DIVERSE PROVOCATIONS, IN BRIEF

When I had written all my main accounts of how to go looking for dynamic mechanisms in specific morphogenetic events, of course I had a collection of pieces left over; and like a good museum curator, I had to find room for them. But I'm not a museum curator; I am presenting history not as an end in itself, but only if I think it provides provocation for future action. To my mind, there is something very stimulating about each of the things I mention here, in relation to thinking about the possibilities of reaction–diffusion mechanisms.

7.3.1 Definitely reaction–diffusion, but not alive: CIMA

Vitalism, in the restricted form that some substances could be made only with the aid of a 'vital force' present only in living matter, is most generally considered to have died in 1828, when Friedrich Wöhler synthesized urea in the laboratory, and 'organic' chemistry thereby acquired a less mystical significance. But is the sauce for the gander sauce also for the goose? If one finds in an inanimate chemical system some dynamics of pattern formation that are unequivocally Turing-type reaction–diffusion, should one adopt a 'chemicalism' that denies the probable occurrence of the same kind of dynamics in living material?

It was also in 1828 that an oscillating chemical reaction was first observed, according to Degn (1972). In more recent times, what has become the most famous oscillating reaction was reported by Belousov (1959) and Zhabotinskii (1964). Both initial publications were in Russian, and attracted little attention in the West until the 1970s. The Belousov–Zhabotinskii reaction is oxidation of a small organic molecule, malonic acid, $CH_2(COOH)_2$, by an inorganic oxidant, BrO_3^-, with the cerium ion Ce^{4+} as catalyst, in aqueous solution. And because all reactants and intermediates are small molecules and ions in water, they all have roughly the same diffusivity, which forbids the generation of Turing patterns. Nevertheless, although this system is often set up to give oscillatory behaviour in a well-stirred system with no possibility of showing spatial patterning, it is also well known for producing travelling waves, planar and parallel in a cylindrical, test-tube-like reactor, or emanating spirally from a number of centres in a Petri dish geometry. Despite the inability of this system to produce stationary patterning, it attracted sufficient attention in relation to life to be mentioned in an undergraduate-level physical chemistry text (Moore 1972), with references to Turing (1952), Prigogine (1967) and Glansdorff and Prigogine (1971), and with the question: 'But is this not exactly what we mean by a living system – a localized region of order that maintains itself by feeding on the free energy stores of its environment?'

Stationary pattern was found, however, as soon as a system of this sort was set up to react in a polyacrylamide gel reactor that, by the presence of a rigid framework to which some molecules could become temporarily attached, provided the possibility of different substances having different diffusivities in what was otherwise still an aqueous solution. The CIMA reaction (chlorite, ClO_2^-; iodide, I^-; malonic acid, $CH_2(COOH)_2$), in such a reactor and with a starch indicator to make iodine visible, produced striped or spotted (hexagonal array) patterns for different compositions of the reaction mixture (Castets et al. 1990; Ouyang and Swinney 1991). These authors reported these as Turing patterns. Lengyel and Epstein (1991) devised a chemical mechanism for the reaction, having Turing dynamics with the morphogens X and Y identified as I^- and ClO_2^- respectively.

These reports claimed the CIMA reaction as the first clear experimental evidence of Turing patterns. This, I believe, is true, if 'clear evidence' must include the chemical identity of the morphogens. But if Turing behaviour was obtained so readily by adding to an aqueous system what is needed to produce a range of diffusivities, I think it would be quite astonishing if conditions for Turing patterning do not arise commonly in the chemical systems of life. The great difficulty, however, is to identify the morphogens, amidst all the complexity of a living system. To my mind, dynamic evidence, short of that identification, can be conclusive. That has seemed to be accepted for the observations of Kondo and Asai (1995) on angelfish patterning, which I describe in Section 9.2; and throughout this book I seek to stress the significance of evidence on the dynamics of patterning processes.

7.3.2 Polar coordinates

The surveyor must measure distances and directions; but there is more than one geometrical formalism that can be used to represent the results of such measurements. The Self-Organizer must use the thing being measured as also the medium for recording the results of the measurements. In this book and earlier (Harrison 1993), I have stressed the quantitative aspect of setting up a scale of distance along a straight line. But such a line can be a part of different coordinate systems. In 2-D, reaction–diffusion theories are very commonly studied on flat rectangular domains, using 2-D Cartesian coordinates. For the flat disc, with formation of whorled structures, a polar coordinate system is the obvious choice (though neither I, Section 3.2, nor Lacalli (1981), actually had to write down a symbol for an angular coordinate in reporting our work). For the hemispherical shell (Section 4.4), however, I did use spherical polar coordinates with angles θ and φ. Kauffman et al. (1978) approximated the wing imaginal disc of Drosophila (from which the wing develops at metamorphosis) as an ellipse, and used elliptical coordinates. While thinking about patterning in the Drosophila embryo (Chapter 8), I have struggled, so far to no great effect, with confocal ellipsoidal coordinates, somewhat familiar to chemists in the quantum mechanics of the hydrogen molecule-ion.

But there are certain phenomena of development that, for the 'generalization from data' stage of the scientific method, strongly demand the use of a circular polar coordinate, usually together with a linear coordinate perpendicular to the plane of the circle to make a cylindrical coordinate system. These events have the striking feature

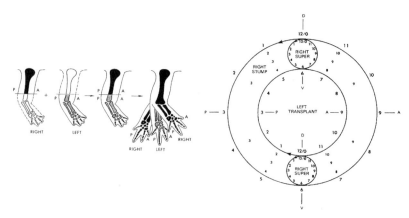

Figure 7.6 The result of one type of limb transplant experiment in a newt (the phenomenon occurs essentially similarly in insects) described in terms of a circular coordinate represented in clockface manner, clockwise for a right limb and anticlockwise for a left limb. Left transplant, rotated 180°, onto right stump leads to growth of two supernumerary limbs, both right. From French *et al.* (1976) with permission.

that they are in enormously diverse organisms: limb morphogenesis in insects, e.g. the cockroach; the same in newts (for both, French *et al.* 1976); and reversals of chiral asymmetry in the shapes of the unicellular ciliates (*Tetrahymena* and, best known in elementary biology courses, *Paramecium* and *Stentor*; Frankel (1989)). Fig. 7.6 shows the result of one of many varieties of limb transplantation experiment that can be done both in insects and newts: in this case, part of a right limb is amputated, and the missing part replaced with a graft from a left limb, rotated 180° from the correct orientation. Two supernumerary limbs arise from the join; both are right limbs. Positional values around the limbs are assigned in an angular sense, as clockface numbers up to 12. The predictive rule for development after excisions or transplants is: inter-calation of missing values, and supernumerary limb formation where a whole circle of values is missing. This illustrates very well that a gener-alization from data can be powerfully predictive even when it does not go beyond that level and connect the data to any putative mechanism. Frankel leans towards reaction–diffusion as a possible mechanism for the cylindrical coordinate property of the ciliates; but he also has important generalizations in dividing his accounts of morphology and development in several species into 'statics' and 'dynamics', and in recognizing the cylindrical coordinate model as a statement about topology of the shapes of the organisms (a point which I make about gastrulation in Section 9.4).

As Fig. 7.6 illustrates, problems requiring the use of a cylindrical coordinate system involve the property of handedness (chirality). At the end of the Preface to this book, I raised the question of how L-aminoacids only can make both left and right hands, a problem I have earlier called 'big hands from little hands' (Harrison 1979). More precisely, especially in regard to the hunt for mechanisms, the question may be put: going from molecules upwards, at what spatial scale does living organization change from making only one hand to making both? Protein monomers will self-assemble, i.e. fit together, in large numbers to make structures that are still homochiral (single-handed): e.g. cytoskeletal components such as microtubules and actin microfilaments; and viruses, such as the tobacco mosaic virus in Fig. 1.1. But my question has a known observational answer, from the organisms described in Frankel (1989). A ciliate cell has a chirally asymmetric surface (particularly, a twisted mouth leering to the left or the right) made up largely of assemblies of microtubules. It is possible to generate mirror-image joined pairs of cells, and even individual cells of either chirality on that macroscopic scale. The size range of ciliate cells is roughly $10-100$ μm.

7.3.3 Identities of Turing morphogens?

The evidence most commonly given against reaction–diffusion as 'the chemical basis of morphogenesis', Turing's supremely confident assertion as the title of his paper, is the general lack of unequivocal identification of activator and inhibitor morphogens in the Turing sense. I have made it clear that I advocate a more general concept of the term 'Turing dynamics', as applied to the pattern-forming abilities of mechanisms combining activation, inhibition and communication, than one requiring the protagonists to be two molecular species. Nevertheless, I think they often will be, so I list here some of the more promising candidates for the style and title of Turing morphogen:

(1) In an inanimate chemical system, I^- and ClO_2^- (CIMA reaction, Section 7.3.1).

(2) In the slime mould *Dictyostelium discoideum*, in which an aggregate of cells undergoes patterned differentiation of two types of cell, Gross *et al.* (1988) proposed an organic molecule known as DIF-1 (because it is known to take part in the

differentiation event) as activator morphogen X, and ammonia, NH_3, as inhibitor Y because of their roles in controlling a chloride channel and a proton pump in the wall of a calcium-sequestering vesicle. (See Harrison 1993, Fig. 10.8).[2]

(3) In vertebrate embryogenesis, some or all of the protein gene products sonic hedgehog (Shh), bone morphogenetic proteins (BMPs), fibroblast growth factors (Fgfs), transforming growth factor (Tgfβ), and substances in the Wnt signalling pathway have been indicated as possible morphogens in various events: lung branching (Section 9.5.2); limb morphogenesis (Section 9.5.1); gastrulation (Section 9.4); chick feather follicle formation; and, of course, mouse hair follicle formation, now indicated as definitely a Turing mechanism with WNT and DKK as the activator–inhibitor pair (Section 6.3.5).

(4) In *Acetabularia*, following a suggestion of Lisman (1985) that autophosphorylating protein kinases that work intermolecularly could make a bistable system, I suggested (Harrison *et al.* 1988, 1997) that an integral membrane protein of this sort, activated by binding extracellular calcium, might be an X morphogen in whorl formation (Fig. 3.2).

(5) In plant apical meristems, especially *Arabidopsis*, gene products are being found that could be involved in Turing mechanisms, at least as I read the published information about them. Flower development genes intermediate between meristem identity and organ identity genes (*FIM*, Simon *et al.* 1994; *UFO* (unusual floral organs), Ingram *et al.* 1995) in the *Arabidopsis* shoot apical meristem, extracellular protein CLV3 and the membrane protein to which it binds, a kinase CLV1 (Reddy and Meyerowitz 2005).

(6) In trichome patterning on plant leaves, the proteins GLABRA (activator) and TRIPTYCHON (inhibitor) (see Digiuni *et al.* (2008)).

---

[2] See also Palsson and Cox (1997) and Palsson (2007) for further modelling work on *Dictyostelium*.

Part III   But animals are different

By far the greater part of biological science, quantitatively, is devoted to animals, because we are among them, and more specifically because the study of them may advance medicine. There are two great differences between animals and plants from the viewpoint of how they develop. First, animals have a precise body plan, the formation of which must be completely executed before they can properly function. Second, in their developmental processes, cells move, changing their relative positions sometimes individually and sometimes by concerted streaming of large numbers of cells together.

Both of these differences make experimental pursuit of the concepts of pattern formation more difficult in animals than in plants. The constancy of numbers of parts, such as limbs, well regulated in the face of varying embryo size, is a challenge to theory, because it shows that developmental mechanisms can do more than mark out a scale of distances, but can also adjust that scale to embryo size. For plants, I have shown by a number of examples in Part I that quantitative spatial measurements during development are very valuable in trying to correlate theory and experiment. I discuss spatial measurements in relation to *Drosophila* segmentation in Chapter 8; but I think that any detailed account of the phenomenon of gastrulation, which I discuss briefly in Section 9.4, will show the reader that it would be very difficult to devise a programme of quantitative spatial measurement to study the applicability of particular theories there.

I have not myself done experimental work on animal development. For two phenomena, localization of the developing heart in a salamander and the beginnings of segmentation pattern in a fruit fly, I and my group have been involved in substantial theoretical work related to detailed experimental information. But a number of the phenomena I touch on, such as gastrulation and somitogenesis, are ones I have not worked on, and are described more briefly and with fewer illustrations. In knowledge of the phenomenology and hierarchical ordering of developmental events, biologists will be far ahead of me. Physical scientists lacking this background can find a very accessible general account of animal embryology in Slack (1983). He also shares my liking for what he describes as 'dynamical systems theory'.

8

The dreaded fruit fly

Why 'dreaded' and by whom? There is more than just an urge to be facetious in my title for this chapter. I have two reasons to dread the fruit fly. First, when I, and other practitioners of dynamic theories of development of my acquaintance, have tried to make some definite contact with experimental biologists, we have of course received diverse reactions from different individuals. But there have also been group characters discernible for people studying particular kinds of organism. And among the groups that I have had contact with, the very large one devoted to the fruit fly *Drosophila melanogaster* has seemed, particularly through the 1990s, to be one of the most difficult with which to sustain an interaction. There is, I think, a very sound reason for this, and it is much more sociological than scientific. Through the past half-century, the dynamic theory group has remained quite small, while the fruit fly group started from small beginnings and exploded to many thousands of workers, mainly seeking to identify new genes through their functions as revealed by mutations. That highly competitive and attention-absorbing field tended, I believe, to widen the gap in approaches and attitudes between molecular biologists and macroscopic-scale dynamic theorists that, for the proper advance of developmental biology, should be a shrinking gap. Upon giving a talk on the possible application of reaction–diffusion concepts to the earliest events in *Drosophila* segmentation, I have experienced a prominent geneticist in the field jumping up to say: 'But you don't need any of this.' And I have heard another geneticist describe another reaction–diffusion theorist's talk as 'pathetic', both being well-established people in their fields. Dislike for this kind of theory can be peculiarly intense, and I have found that attitude most specifically in relation to *Drosophila*. Nevertheless, if one takes a perspective on the literature through the 1980s and onwards,

163

one can find a strong thread of interest in the chemical-dynamic possibilities running through it. I describe some of this in Section 8.2.

Second: my first point is about interactions with people; the second is about what they have written. A news feature in *Nature* (Knight 2003) starts with this quotation from Francis Crick: 'There is no form of prose more difficult to understand and more tedious to read than the average scientific paper.' Knight goes on: 'It wasn't always so. Crick and others of his generation, who began writing scientific papers in the 1940s, have witnessed the transformation of scientific prose. A form that was as readable as the average newspaper has, in some fields, become a jungle of jargon that even those familiar with the territory struggle to understand.' I started reading and writing scientific papers at the start of the 1950s, and I quite agree. The fields that have suffered most are, I think, those that have grown most rapidly, so that people become totally competitively absorbed in their specialty. *Drosophila* genetics is surely one of these. For example, it is essential to my thinking about segmentation mechanisms (Section 8.2) to know whether anything proposed from detailed genetic studies involves longitudinal (antero-posterior) interactions, or whether it does not; and I often have difficulty discerning how the authors think about this from a published paper.

The above rather negative comments are intended to point towards directions for progress, not to offend anybody. Since the beginning of the millennium, I have begun to find it possible to initiate collaborations with people in this field!

8.1 BODY PLANNING

8.1.1 Animal plans and plant plans

Given that I presume to discuss animals when I have already devoted a large fraction of this book to plants, it must be evident that I have some expectation that what is discovered about development in the one kingdom will illuminate what is going on in the other. There are, however, three substantial differences between developing plants and animals that could lead one to doubt that expectation. First, cells in plants, once formed in division processes, do not later change their relative positions, whereas such motion (e.g. 'cell sorting' in instances in which groups of like cells tend to group together) is a common and significant feature of animal development. (To mention this is not trivial. I have encountered erudite and broad-minded people who had spent years immersed in studies of animals, and admitted to quite a shock on discovering the absence of relative motion of cells in plants.) Second, plants generally continue to develop throughout their lives, while animals generally reach a fully developed form that is then maintained essentially unchanged for a large part of the life span. Third, that fully developed form exhibits a 'body plan' that is the same for all individuals of the species. Two dogs must each have four legs. Two oak trees do not have to have the same number of branches.

The formation of body plans is therefore one of the major pre-occupations of animal developmental biology. Must its mechanisms have nothing in common with those of plant development? Here, I believe we need careful definition of what is to be compared with what. A plant is not, developmentally, an individual in the same sense that an animal is. As discussed extensively in Part I, a plant commonly has numerous regions, each growing and forming patterns, but each developmentally isolated from all the others. If each of these regions is regarded as an individual organism, the correspondence between plants and animals becomes much closer. As any dog has four legs, any wild rose has five petals; and many flowers have lengths of stamens and pistil and sometimes elaborate shaping of petals precisely designed for insect or wind pollination. Can such a region be thought of as a body, with a body plan having features as precise as that of an animal? From the viewpoint of comparing computations of develop-ment with the natural phenomenon, I often worry that, in the midst of vast accumulations of knowledge about mutations, comparatively little work is done on the variations (especially quantitative ones, of numbers and distances) in a wild-type population. This topic is dis-cussed for *Drosophila* in Section 8.3. But to see variability in a population which is actually clonal, simply look at all the flowers on a single rhododendron tree.

8.1.2 'Fushi'

A feature of special importance in many animal body plans is segmentation. It is a defining characteristic for the class Insecta, and is essential – though not so externally obvious – in the vertebrates. A *Drosophila* gene concerned in the earliest appearance of segmental division of the body is called by the Japanese name *fushi tarazu*, translated as 'not enough segments'. This describes the nature of the defect in an embryo in which this gene is deficient in its operation. But the original meaning of *fushi* is the segmentation of a bamboo stalk, and the word can also be used metaphorically for such concepts as rhythm. The breadth of meaning is not trivial: it points to the values both of comparing animals and plants, and comparing spatial and temporal periodicities. As will emerge in what follows, the first marking-out of the segmental pattern in *Drosophila* is essentially simultaneous, but the segments appear sequentially in many insects; and for the kinds of possible pattern-forming mechanisms that I explore, the change from simultaneous to sequential can be made by a change in the value of a parameter, or the imposition of a shallow gradient along a system. The latter was illustrated for a Brusselator-type mechanism in Lacalli *et al.* (1988) and a cover picture accompanying Harrison and Tan (1988), both from the same computation by Lacalli (see Fig. 8.5).

The ability of living material to form repeats of itself or of parts within itself, on a temporal or spatial basis, is close to providing a basis for a definition of life itself; segmentation is something that the Self-Organizer has an overwhelming urge to do. Does this generalization state the only common basis we shall ever find for segmentation events, or should we be looking for a level of mechanism in which the segmentation of all insects is the same, and at which the mechanism defines not life itself, nor one genus or species, but why there is such a thing as the class Insecta. Davis and Patel (1999) tackled this topic for segmentation in arthropods, annelids and chordates, starting with a statement that 'within phyla ... the homology of segments is generally accepted', and finishing 'it is perhaps too soon to conclude that segmentation is homologous between the various phyla. Indeed, the opposite conclusion of convergence at the level of developmental mechanism is perhaps more intriguing.'

That paper, and the whole topic in general, clearly raises by implication the semantic questions of what different kinds of scientists mean when they use the term 'mechanism' and what various biologists understand by the term 'homology' of structures in different groups of organisms. My use of 'mechanism' for the particular concepts that I pursue is that of the physical chemist, in which a mechanism is first a list of chemical equations for unit steps in a reaction scheme and then the corresponding set of rate equations for changes in concentrations of substances. But in Section 4.2 I have introduced also the physicists' or mechanical engineers' concepts of mechanical deformations as mechanisms. What are we to seek as 'mechanism' when on the one hand we would like to find something that all Insecta have in common to generate a segmented body, and on the other hand we have available a vast amount of detail on the gene-expression hierarchies in the development of *D. melanogaster*? The Diptera, to which they belong, are among the most highly evolved insects, and may have acquired some of those details by applying to a simpler ancient version the German adage 'Warum so einfach; es konnte compliziert sein?' (Why so simple; it could be complicated?)

8.2 THE EARLIEST APPEARANCE OF SEGMENTATION IN *DROSOPHILA*

The kind of theory I am exploring in this book, starting from Turing (1952), preceded by about three decades the experimental discovery of the earliest appearance of segmentation in *D. melanogaster* (Hafen *et al.* 1984). It is pertinent to ask what Turing and subsequent elaborators of his kind of theory predicted, and whether later experiment correlated with such predictions. Turing's title, 'The chemical basis of morphogenesis', essentially stated that patterning of chemical concentration distributions should be found to precede their expression in change of shape (or differentiation state). Following Prigogine's (1967) devising of the Brusselator scheme, Herschkowitz-Kaufman (1975) published a long mathematical biology paper exploring some of the Brusselator's properties. One of these was that pattern formation could be confined to only a part of the available region by a spatial gradient in the input concentration A, together with uniform concentration of the other input B. She showed two patterns of repeated parts for the same A gradient and different values of B (Fig. 8.1). She could have chosen other values for B, so it is quite fortuitous that the pattern of seven wavelengths occupying about two-thirds of the length of the system is astonishingly close to the earliest segmentation pattern in *D. melanogaster* (Fig. 8.2C). But the general correspondence to Turing's prediction of chemical patterning preceding morphological change is quite clear.

The sequence of expression of three types of gene shown in Fig. 8.2 is now one of the best-known illustrations in developmental biology. But physicists and chemists erudite in their own fields are often completely unacquainted with it, so it needs explanation. The *Drosophila* egg is about 0.5 mm long (give or take about 30% variation in length in a population of eggs) and approximates a prolate ellipsoid of revolution about the anterior–posterior axis (somewhat flattened on the dorsal side). In the oviduct, it is surrounded by 'nurse cells', which inject into the egg the mRNAs from some maternal genes that have been transcribed in the mother. These usually remain localized near

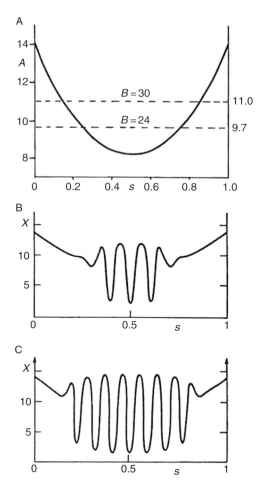

Figure 8.1 Restriction of a spatially repeating pattern to a part of the length of a system by smooth continuous variations in the values of input concentrations, from Herschkowitz-Kaufman (1975) with permission. Brusselator mechanism with reactant input concentrations *A* and *B*. *s* is distance along the system. *A* is distributed in a double-ended exponential (i.e. hyperbolic cosine) gradient. *B* has a constant value along the system of 30 in B and 24 in C.

one or other of the ends of the egg, occupying only a small percentage of its length. Translation yields the corresponding proteins, which are mobile. The system therefore has the prerequisites for establishment of gradients of the proteins of exponential form with respect to distance along the embryo, as discussed in Section 6.4. For the Bicoid (Bcd) protein in *Drosophila*, such a gradient, of quite precisely exponential

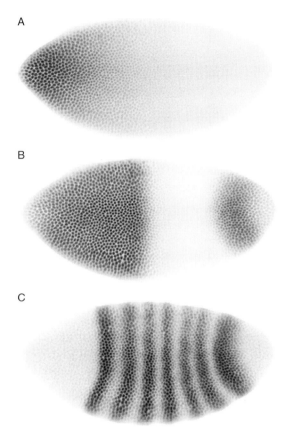

Figure 8.2 Protein concentration distributions, i.e. gene expression patterns, in a *Drosophila* egg at interphase 14, i.e. when it contains about 8000 nuclei, all in a single cell and forming a single layer (syncytial blastoderm) just inside the surface of the egg, which is roughly a prolate ellipsoid of revolution 0.5 mm long. Anterior–posterior direction is left–right. The small patches are individual nuclei, and dark is high concentration. A: Bicoid (Bcd) protein, a 'maternal-effect' gene: DNA is transcribed into mRNA in the mother, and the mRNA is deposited in the anterior region of the egg and there translated into the protein. B: hunchback (Hb) protein, from the egg's DNA. C: even-skipped (Eve) protein, from the egg's DNA, and one of several genes similarly expressed in a seven-stripe pattern that is the first visible sign of the segmentation that makes the organism an insect. Image from the FlyEx database (http:// urchin.spbcas.ru/flyex/), see Poustelnikova *et al.* (2004).

form, has been observed along about 80% of embryo length (Driever and Nüsslein-Volhard 1988; Houchmandzadeh *et al.* 2002; Holloway *et al.* 2006; Fig. 8.3). It seems to be generally accepted by biologists that

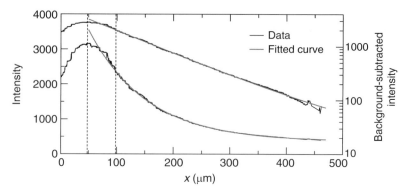

Figure 8.3 For a Bcd pattern like that in Fig. 8.2A, plots of concentration and logarithm of concentration versus distance along embryo showing that the gradient is exponential along about 80% of embryo length. See Section 6.4 for the explanation of an exponential gradient by diffusion and first-order decay. From Houchmandzadeh *et al.* (2002), with permission.

exponential gradients exist and are explicable by local sources of a diffusible substance that also decays.[1] This is assuredly a reaction–diffusion mechanism, but is not commonly referred to as such. That term tends to be restricted to the formation of more complex pattern than a monotonic gradient.

Beyond the 'maternal-effect' genes such as *bicoid*, we need to consider where the embryo's own (i.e. 'zygotic') nuclei are. From the single fertilization nucleus, 13 successive divisions, occurring close to synchronously in all daughter nuclei, generate about 8000 nuclei. These have progressively moved outwards until they form a single layer just inside the surface of the egg. All of this happens in four hours, and all the nuclei are then still within a single cell or syncytium. This is the syncytial blastoderm. Then, at interphase 14, there is a pause of one hour in the division sequence; and by late in that hour, membranes have formed to enclose each nucleus in a separate cell, forming the cellular blastoderm. But before that happens, gene expression patterns like those shown in Fig. 8.2B and Fig. 8.2C arise. Fig. 8.2B is the pattern of expression of the Hunchback (Hb) protein, an example of the class of genes known as 'gap' genes. A null mutation in one of these genes leads to large gaps in the patterns of subsequently expressed genes and in the

---

[1] See Gregor *et al.* (2005, 2007a, 2007b); Weil *et al.* (2006, 2008); Coppey *et al.* (2007, 2008); and Spirov *et al.* (2009) for recent developments in the dynamics of *bcd* gradient formation.

morphology of the insect. Fig. 8.2C is the strikingly periodic-looking expression pattern of the Even-skipped (Eve) protein, one of the class of genes described as 'pair-rule', because for each gene that is expressed as one set of stripes there is another that forms stripes in the spaces in that set. (For Eve, the gene that thus dovetails in is Ftz (*fushi tarazu*). And for readers unfamiliar with genetic notation, *eve* is a gene, i.e. it is DNA, and Eve is its translation product, i.e. it is a protein. The convention is different in plants; to botanists, a gene is in italic capitals, e.g. a gene called unusual floral organs is *UFO*.) Beyond these three classes of genes there are of course many others, the next being 'segment polarity' genes (e.g. *engrailed*) that specify for each segment which end of it is anterior or posterior.

But the present account is concerned only with the first three classes. These three, illustrated in Fig. 8.2 by *bcd, hb* and *eve* are hierarchically linked. Mutations in an upper level of the hierarchy affect what happens at all lower levels. And it really looks as if the Self-Organizer has over-reached himself in a Byzantine excess of over-organization by using such a strange sequence of patterns to arrive at the apparently simple periodicity of the Eve pattern, which, along with all the other pair-rule patterns, is the primary layout that is to become the segmentation pattern of the insect. It looks periodic, and can legitimately be described as periodic just as a description of its shape. But does that inevitably imply its formation by a mechanism for generating spatial periodicity, i.e. a waveforming mechanism? By way of example, Akam (1987) wrote a review of the molecular basis for this patterning in which he indicated distinct promise for Turing-type mechanisms. (I visited him when he was writing it, and know that he was quite interested in this possibility.) But only two years later, Akam (1989) wrote a 'News and views' item in *Nature* entitled 'Making stripes inelegantly', in which he presented evidence for each stripe being formed under individual control without an overall periodic pattern for all seven stripes.

Such evidence, largely in the form of the discovery of a multitude of transcriptional regulators for most of the genes, has led to a divergence of views on the probable general nature of the patterning mechanisms. There are two crucial and interrelated questions.

First, when there is a hierarchical relationship between classes of genes, like those for which Fig. 8.2 shows spatial distributions of expression, is there also an aspect of 'downstream autonomy' in the patterning of each gene, or is the hierarchy everything?

Second, does each nucleus respond only to local information, or is there 'crosstalk' along the anterior–posterior direction of the embryo?

These questions are related to two different meanings of the word 'morphogen', to which I devoted a chapter in Harrison (1993), proposing the following terminology.

Type I morphogen: a substance distributed in a monotonic gradient (linear, exponential or whatever), the concentration of which locally informs cells, nuclei or other substances how to develop. This corresponds to the concept of 'positional information', for which the botanist Hermann Vöchting (1877, 1878) has priority, and which has been promulgated strongly by Lewis Wolpert (1970, 1981, 2002), mostly with reference to animals. This is the common usage of 'morphogen' among biologists today. And the one-morphogen, local source, diffusion, decay mechanism for the formation of a steady-state exponential gradient is the kind of reaction–diffusion mechanism widely accepted among biologists.

Type II morphogens: substances participating in pattern formation mechanisms in the general Turing class, capable of forming spatially complex concentration distributions that are not monotonic. Such substances interact in pairs – hence I define them in the plural. This is the usage of 'morphogens' among chemical-kinetic theoreticians. The *Oxford English Dictionary* gives both meanings, and accords Turing the priority for inventing the word 'morphogen'.

For the *Drosophila* segmentation gene hierarchy, the positional information concept starts from the Bcd exponential gradient as a Wolpert signaller or type I morphogen. If we take it for granted that this gradient exists and that there is a ready explanation of its exponential form, we may proceed to consider the reading of its signals and whatever else may be necessary to generate the gap gene and pair-rule gene patterns. For these, and most especially the latter, my questions regarding 'downstream autonomy' and existence or not of 'crosstalk' are very relevant. To my understanding, all of the vast amount of information that has been accumulated about details of gene interactions has not yet produced a definitive model of the formation of the periodic seven-stripe patterns and has not yet answered my two questions. Notwithstanding divergent philosophical views on whether developmental biology will ever acquire all-embracing theories of how development happens or whether it will have to continue to be a compendium of descriptions of immense diversity, I feel that those two questions should have achievable answers in any particular example. Yet they remain open questions for the *Drosophila* segmentation pattern, despite the vast amount of information that has been acquired about it. To my mind, the most balanced review of this topic is that of Pick (1998). She points out that, despite the evidence for single-stripe control of *eve* stripe 2, 'single stripe elements have not been found for all of the *eve* stripes. Therefore, it is not clear yet whether the elegant *eve* stripe 2 story is the rule ... or whether this story is the exception among more complex regulatory modes.' She goes on to indicate that for *ftz* and *paired*, 'the striped patterns are set up by modular regulatory elements that direct expression in seven stripes' and concludes that 'it is not yet possible to provide a mechanistic explanation of a complete striped pattern for any single pair-rule gene. Clearly, much remains to be learned about how periodicity is generated, even in *Drosophila*.'[2]

The local gradient-reading concept starting from Bcd has been extensively elaborated in various ways. It somewhat strains credulity that this gradient, so different in concentration level and slope at its anterior and posterior ends, should be the omnipotent commander of the production of seven evenly spaced stripes. But there are also maternal-effect genes, e.g. *nanos*, that have their mRNAs injected at the posterior end of the egg and produce posterior-high protein gradients (e.g. Irish *et al.* 1989) (Fig. 8.4A, broken line). It is possible to devise mechanisms for gap and pair-rule pattern formation based on local reading of the concentration ratios of opposing and crossed gradients (e.g. Pankratz *et al.* 1990). A more complex model in this category can be found in Hartmann *et al.* (1994). For example, stripes 3 and 4 of the pair-rule gene *hairy* form as a single broad stripe; the gradient ratio idea shows how that domain can be split into the two *hairy* stripes with room for the dovetailing *runt* stripe between them. Fig. 8.4B and C give an indication of how the spatial domains of expression of five gap genes correspond to the positions of the *eve* stripes.[3]

The alternative possibility that upstream genes provide reactants like, e.g. Brusselator A and B for the downstream genes to produce pattern semi-autonomously by something perhaps like reaction–diffusion dynamics, but definitely needing crosstalk along the antero-posterior direction is of course the kind of mechanism I lean towards (Harrison and Tan 1988; Lacalli *et al.* 1988). To set up a model of this kind using 1-D modelling, I assumed that a double-ended A gradient like that of Herschkowitz-Kaufman (1975) (Fig. 8.1) could represent the combined

---

[2] For a recent review focused on the regulatory elements involved in stripe formation, see Papatsenko (2009). Clyde *et al.* (2003) is also a very good demonstration of multiple stripe control.

[3] But the gradient ratio idea does not address the effects of variability in the upstream gradients, and the potential for such variability to destroy stripe formation. See Holloway and Harrison (1999b), Manu *et al.* (2009a) for further discussion.

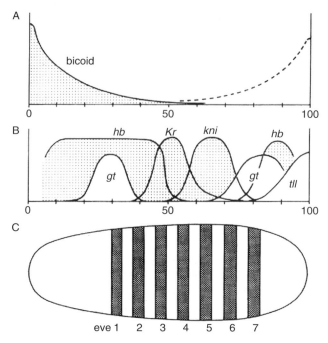

Figure 8.4 Schematic (designed by Lacalli) of patterned expression of gene types: A: maternal-effect; B: gap; and C: pair-rule at the times of their hierarchical interactions (early in interphase 14; all of these patterns are time-dependent – see, e.g. Jaeger *et al.* (2004a, 2004b) and Surkova *et al.* (2008)). Abscissae numbered 0–100 are distance along the embryo from the anterior end in percentage of embryo length. Ordinates in A and B are concentration of protein. The broken line in A shows a posterior-high exponential gradient, such as the gene product *nanos*. The indications for Bcd, Hb and Eve are from data like those shown in Figs 8.2 and 8.3. The two gradients shown in A might together be equivalent to the A gradient in Fig. 8.1A. B and C raise the question of whether it is reasonable to postulate the segmentation pattern as achieved by reading of gradients, or ratios of gradients, in the gap gene patterns. Those genes are: Hb, Hunchback; Gt, Giant; Kr, Krüppel; Kni, Knirps; Tll, Tailless. From Lacalli and Harrison (1991) with permission. (See also Gilbert (2006).)

effect of anterior-high and posterior-high gradients of two different proteins. In 1990 I and my group could only manage to get such ideas into publication by using a physics journal (Lyons *et al.* 1990). I wrote (Section 10.2 of Harrison 1993):

> Is it reasonable to set up a model by rolling the effects of two or more different substances together into a double-ended gradient of a single substance A? I have been severely criticized by drosophilologists for doing

this, and the procedure certainly runs quite counter to their energetic efforts to disclose the variety of detail in the actions of several substances forming the anterior and posterior gradients.

The contrast here is that the 'mainstream' method is to look for the structural chemical nature of all the substances involved in any way in the mechanism; to anyone seeking a dynamic explanation, two substances are effectively the same if they are both able to produce downstream effects on the same time-scale, even if they differ substantially in structure. A substance in the mechanism is characterized by its kinetic properties in association with the other substances involved, not in its molecular structure.[4]

Unlike the simple 1-D gradient-reading mechanisms, 'crosstalk' mechanisms need to take into account that the garden fence goes all around the property, and enquire why repeating pattern happens only in the antero-posterior direction, not also at right-angles to it, making spotted rather than striped patterns. The matter of stripes-versus-spots is much more general than the *Drosophila* topic. For that particular case, however, Lacalli tackled the formation of stripes in two ways: first (Lacalli *et al.* 1988) by showing computationally that a Brusselator-type mechanism, which most commonly tends to make spots, can be forced into striping if it is hierarchically dependent on a 1-D gradient (of the rate constant c of the autocatalytic step, (Eq. 5.1c)). Later (Lacalli 1990), he showed that a four-morphogen model of two symmetrically linked side-by-side Brusselators would make stripes. This model is in fact in a general symmetry class that Lyons and Harrison (1992a) showed to be a stripe-maker, first in relation to reaction–diffusion but then (Appendix of the 1992 paper) in a quite different kind of model of mutual reinforcements and mutual inhibitions in synapses in the brain (see Section 9.2).

What of the comparative perspective on segmentation mechanisms, not going as far as spanning vertebrates, annelids and arthropods, nor yet all the arthropods, but only various cases in the class Insecta? I have already mentioned that some insects form their stripes sequentially. Roughly, the distinction is that 'long germ band' insects – such as *Drosophila* – in which the stripes take up most of the length of the egg, form them more or less simultaneously, while 'short germ band' insects, in which the stripes, and hence what is to become the whole insect, occupy a smaller fraction of the egg length, make the stripes sequentially. Among these latter, studies on a beetle, *Tribolium* (Wolff

---

[4] For other recent papers on double-ended gradients in *Drosophila*, see Houchmandzadeh *et al.* (2005), and Howard and Rein ten Wolde (2005).

et al. 1995), and a grasshopper, *Schistocerca* (Davis *et al.* 2001), have shown expression in early segmentation of some pair-rule genes, but by no means the whole suite of them that seems so important in *Drosophila*. Peel (2004) reviewed the evolution of arthropod segmentation mechanisms. Despite a focus largely on these few insects, he was led into a comparison with vertebrate segmentation, which is always sequential, first appearing in the successive formation of small boxes of cells called 'somites'. These are essential to the later formation of the vertebral column, the nervous system and the musculature. Somite formation involves a timing mechanism called the 'Notch signalling-dependent segmentation clock'. Peel's conjecture is that a similar mechanism operated in the earlier insects and was gradually lost in evolution to the Diptera, such as *Drosophila*.[5]

Recently, the accumulation of vast amounts of information on the composition of genomes has begun to lead to the concept of networks of interactions between genes, especially via their protein products. This leads to the possibility of a developmental mechanism having a complexity vastly exceeding that of the structure of the genome itself. But how should one look at a network to try to discern the essentials of simpler mechanisms within it that can generate various pattern-forming events? For *Drosophila* segmentation, a start on this was made a few years before the current flurry of interest in networks by using the concept of 'gene circuits' (Reinitz *et al.* 1995; Reinitz and Sharp 1995). The model is set up for one spatial dimension, and consists of equations for the rates of growth of concentrations of the protein gene products. The terms in the equation for any one protein are: diffusion and decay terms for that protein, and regulatory terms for control of the activity of the gene for that protein by the concentrations of protein products of several other genes, e.g. five of them in the second paper cited, which is for formation of Eve stripes 2–5, with regulatory influences from *bcd* and four gap genes, *hb, Kr, gt* and *kni* (abbreviations of a mixture of English and German words for hunchback, cripple, giant and dwarf).[6] The authors point out that these equations are similar to Turing's equations, with the regulatory terms replacing the reaction terms. They also point out that the Eve stripes, though they form very quickly, do not form exactly simultaneously, but in the order: 'stripe 1 forms first, then stripe 2, next stripes 3 and 7, then 4 and, finally, stripes 5 and 6

---

[5] For more recent developments, see Peel *et al.* (2005); Damen (2007); Pueyo *et al.* (2008); and De Robertis (2008).

[6] For more recent papers on gene circuits in segmentation, see Jaeger *et al.* (2004a, 2004b); Manu *et al.* (2009a, 2009b).

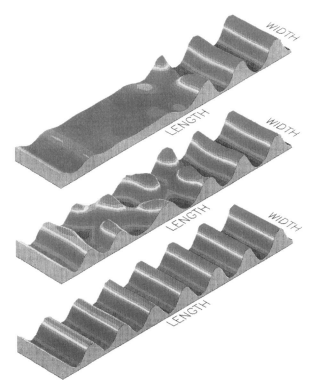

Figure 8.5 Three stages in a computation (times 3000, 6000 and 30 000 arbitrary units) of pattern formation by a variant of the Brusselator mechanism in a fairly narrow rectangular system with right–left gradients in two of the reaction rate constants (high at right end, decreasing by about a factor of 2 along the system). The one-dimensionality of the gradients forces a 'spotting' mechanism to make stripes. (This happened also in a system of the same length but twice as wide.) For relation of order of formation of stripes to those of Eve in *Drosophila*, see text. From Lacalli *et al.* (1988) with permission.

(Frasch *et al.* 1987)'. In their model, early formation of stripes 1 and 2 is related to the placement of the gap genes controlling them. But in Lacalli's calculation with a much simpler model, already cited as used in two cover pictures, one of which is reproduced in Fig. 8.5, the stripes form in the order 1 and 2, then 7, then the others with a slight suggestion of stripe 3 being slightly ahead.[7]

What spatial dimensionality should one use in computational modelling? Just as I illustrated in Part I for plants, at various points in

---

[7] See also Surkova *et al.* (2008) on the timing of stripe formation.

a theoretical project it can be strategic to use 1-D, 2-D or 3-D, and there is no telling in advance which is going to give the most useful information. But all have their specific uses. So far, I have mentioned only 1-D and 2-D work in relation to *Drosophila*. But that more or less ellipsoidal embryo is very 3-D. Questions arise as to whether the domain to be considered for diffusion in the syncytial blastoderm is the thin ellipsoidal shell defined by the nuclear positions, or whether it is the whole volume of the ellipsoid.[8] Either way, the observed precisely exponential shape of the Bcd gradient is a little puzzling, because the derivation of that form (Section 6.4) requires a constant cross-section along the antero-posterior axis. For pattern formation models using Turing-type dynamics in the ellipsoid, to my knowledge fully 3-D source codes have been set up only by Hunding (1993).[9]

8.3 POSITIONAL VARIATION, ERROR AND NOISE IN WILD-TYPE POPULATIONS

So far in this account I have mentioned chiefly what the geneticists have done and what the theoreticians have done, and I may have given the reader a depressing impression of 'two solitudes', a Canadian term for the politics of Québec versus English-speaking Canada; and indeed the analogy of groups using two different languages is appropriate. But in *Drosophila* segmentation work, there are other things going on that provide different approaches to trying to probe the mechanisms. First, as should be evident from my accounts of the work of my group on *Acetabularia* whorls and *Larix* somatic embryos, I feel that much valuable information on pattern-formation mechanisms should be available from spatially quantitative studies on variability within wild-type populations. For *Drosophila* segmentation genes, this is now available for thousands of embryos from work in the Reinitz group (Kosman *et al.* 1997, 1998), and much of it can be found on a website (http://urchin.spbcas.ru/flyex/). This is a vast amount of data, and while some analysis of it has been published (Holloway *et al.* 2003; Spirov and Holloway 2003) or is in the course of publication,[10] the gnomes are going to be digging in this data mine for a long time yet.[11]

[8] For example, see Gregor *et al.* (2007a); Spirov *et al.* (2009).
[9] For 3-D *Drosophila* modelling, see also Coppey *et al.* (2007); Okabe-Oho *et al.* (2009).
[10] See Jaeger *et al.* (2004a, 2004b); Poustelnikova *et al.* (2004); Surkova *et al.* (2008).
[11] There is also a very large expression atlas by the Berkeley Drosophila Transcription Network Project: http://bdtnp.lbl.gov/Fly-Net/; see also Fowlkes *et al.* (2008).

A few simple facts are significant, perhaps in relation to animal body plans in general: eggs in a wild-type population vary by up to about 30% in length. The number of stripes in the pair-rule segmentation patterns never varies except in mutants. It is always seven.

Earlier work on eggs with different numbers of copies of *bcd* (1, 2, 3 or 4; Driever and Nüsslein-Volhard 1988) showed that the seven-stripe pattern can shift somewhat up or down the embryo; and from those data it was already clear that the position of any one feature of a pair-rule pattern does not correlate precisely with a specific quantitative concentration of Bcd protein. The extreme consistency of the number of stripes is a little troubling for reaction–diffusion models; the failure of downstream features to correlate with specific upstream concentrations is bad for direct gradient-reading models.

Lacalli and I (1991) promulgated our worries about the exponential Bcd gradient as master controller of *Drosophila* development, pointing out particularly that the low end of such a gradient is prone to much more serious variations and errors than the high end. This account has served as a take-off point for more recent assessments of what can be found in the current extensive data on numerous embryos (Houchmandzadeh *et al.* 2002; Spirov and Holloway 2003; Holloway *et al.* 2003).[12] In the 1991 paper we did not distinguish clearly between the kinds of variability that might be encountered in two different kinds of experimental comparison: variations seen within one embryo, and variations seen between a number of embryos in a population.[13] Some definitions are in order:

Positional variability can be seen and measured by superposing plots of concentration versus distance along the embryo for the same protein in a number of different embryos.

Distance along the embryo can be measured as a percentage of embryo length or in micrometres. Plots are going to stack differently in those two measures. For instance, plots of a concentration that follows $C = C_0 \exp(-ks)$, where s is distance in micrometres, will all superpose on each other exactly if they all have the same C_0 and the same k in μm^{-1}. But if s has been converted to percentage of embryo length, they will show a spread if the embryos do not all have the same length. An example is discussed below, in relation to Fig. 8.7. I am mentioning things like this in the hope that some readers, particularly those of a physical science or statistical bent, may be provoked into looking at the publicly available

---

[12] See also: Holloway *et al.* (2006); Crauk and Dostatni (2005); Gregor *et al.* (2007b).

[13] For recent work on the precision of gradients, see: Tostevin *et al.* (2007); Bergmann *et al.* (2007); Saunders and Howard (2009a, 2000b).

data and finding correlations or interpretations that others have missed, or misinterpretations that others have made. There are lots of pitfalls around the data mine.

Positional error refers to the concept of positional information in the Wolpert sense. For a cell or nucleus that reads its position from the concentration on a gradient, positional error limits are the error limits on concentration divided by the slope of a concentration–distance plot of the gradient. Positional errors are very bad where the slope is very low.

Noise is random variation in concentrations within one embryo at any position and time, and can be converted into positional error if the substance concerned specifies a 'positional information' gradient. For the low concentrations of biologically active substances, the number of molecules in the vicinity of one nucleus in the blastoderm may be a few hundred to a few thousand.[14] Poisson distributions of such numbers can lead to positional error limits of a few per cent of embryo length, sufficient to make pair-rule stripes run into each other (Holloway and Harrison 1999b).[15] They never do, nor does the pattern show increasing random variation towards the posterior end, as one would expect if position were being read with Poisson error limits from the progressively lower concentration of Bcd towards the posterior.

Houchmandzadeh *et al.* (2002) displayed the variability in Bcd and Hb spatial distributions in stacked plots of data for about 100 embryos (Fig. 8.6). The striking feature is that, in the vicinity of about 0.5 embryo length, the variability in Hb is much less than that in Bcd. This suggests rather strongly that the gap gene Hb is not just passively reading the Bcd concentration locally, but is doing something autonomously that suppresses errors read in the upstream information; and such a downstream mechanism must surely involve antero-posterior crosstalk between nuclei.[16] Stacked plots of seven-stripe Eve distributions show about the same spatial variability as Hb.[17]

Statistical surveys like this can be very informative; but so can comparisons of a pair of disparate individuals. Holloway *et al.* (2006) selected from a wild-type population the longest and shortest embryos, differing in length by 23%, and compared the results of plotting Bcd and Eve protein distributions against antero-posterior distance in

---

[14] Bcd was measured experimentally in this range by Gregor *et al.* (2007b).

[15] For recent work on noise in *Drosophila*, see: Spirov *et al.* (2008); Tkacik *et al.* (2008); Okabe-Oho *et al.* (2009).

[16] Indeed, Manu *et al.* (2009a) and Hardway *et al.* (2008) showed error suppression in gene circuit models of Hb patterning.

[17] See Holloway *et al.* (2006).

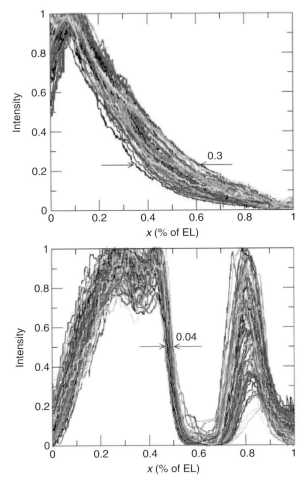

Figure 8.6 Stacked plots for about 100 *Drosophila* embryos of the Bcd and Hb patterns, showing (arrows) that the spread in position (as % embryo length, EL) of the concentration (fluorescence intensity) for which the average position is mid-embryo is much smaller for Hb than for Bcd. From Houchmandzadeh *et al.* (2002) with permission.

micrometres and in percentage of embryo length (Fig. 8.7). The results are not the same for the two gene products. For distances in micrometres, the two Bcd plots are in register, as would be expected from the simple reaction–diffusion mechanism for forming an exponential gradient; but the Eve plots are out of register. Eve patterns come into register when the plots are against percentage of embryo length, as would be expected if the patterns are somehow forming to give the

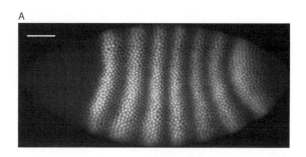

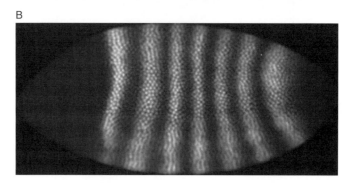

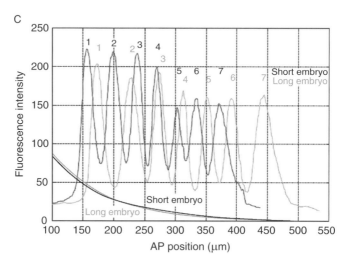

Figure 8.7 Comparison of how spatial distributions of Bcd and Eve change with embryo length, from the shortest and longest embryos in a large wild-type population (A, 468 μm and B, 555 μm) at the same developmental stage. The graphs of fluorescence intensity versus antero-posterior (AP) distance

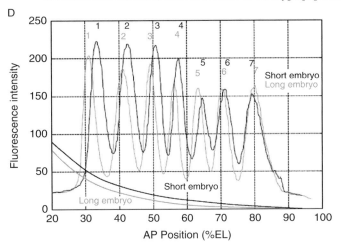

Caption for Figure 8.7 (cont.)

(long embryo, light curve; short embryo, heavy curve) show that: C, the Bcd patterns for the two embryos are in register with each other for distance in micrometres, but, D, out of register for percentage of embryo length (EL). The Eve patterns show the opposite behaviour, in register for the %EL plot. Adapted from Holloway *et al.* (2006), with permission.

same number of segments in the face of variation in embryo length. I intend the word 'somehow' to express a serious challenge to any mechanistic theory, whether it be based on hierarchical reading of upstream gradients or downstream autonomy at the pair-rule level.[18]

From these examples, it should be evident that the study of noise, errors and variability in patterns has a future in regard to the perspectives it can give on mechanisms of pattern formation. In particular, when a mechanism has been set up for computation, it would always be useful to test it against inputs with realistic levels of noise. Holloway and Harrison (1999b) found that reaction–diffusion mechanisms

---

[18] This problem of scaling, how a maternal gradient in micron units can specify a zygotic pattern in percentage of egg length is a strong challenge to classical ideas of positional information via gradient concentrations (e.g. Wolpert (1970, 2002); Kerszberg and Wolpert (2007) acknowledge the challenges Bcd-gap patterning presents). Solutions have been proposed, from pre-patterns in Turing systems (e.g. Figs. 8.1, 8.5; Othmer and Pate (1980); Lacalli *et al.* (1988); Hunding (1989); Lyons *et al.* (1990)) to size feedback on maternal gradient parameters (e.g. Gregor *et al.* (2005, 2008); He *et al.* (2008)).

reading a linear gradient were very efficient at suppressing noise added to the gradient; this had also been shown much earlier by Gierer and Meinhardt (1972), as one little item in a compendious figure showing many properties of their then quite new model.

8.4 EFFECTS OF TEMPERATURE

Reaction rates are notoriously temperature-dependent, unless they are very rapid. The over-used rule of thumb in high school teaching of 'a factor of 2 increase in rate for 10 °C increase in temperature' actually applies rather well to reactions with such rates that their kinetics can be studied in a laboratory class of a couple of hours – and to nothing slower or faster. But this time-scale is not far away from those of events in biological pattern formation; they are often on a scale of minutes and should be somewhat, but not much, less temperature-sensitive than the rule of thumb specifies.

So what about the temperature-sensitivity of patterns formed by reaction–diffusion? In my work on spacing of hairs in *Acetabularia* whorls (Harrison *et al.* 1981), one of the indications of a possible reaction–diffusion mechanism was that the spacing decreased as the temperature was raised, to be expected if the spacing is proportional to $(D / k)^{1/2}$, where the rate constant k has a larger activation energy than the diffusivity D, hence giving the spacing a negative apparent activation energy, $(1/2)(E_D - E_k)$. That was for a case in which the number of parts in the pattern was also quite variable. What about the constancy of the *Drosophila* segmentation pattern?

Not many studies have yet been made of temperature effects in that system. But what has been done is fascinating, and shows once again that Bcd can be as sloppy as it likes; it's the workers down below who are meticulously making a correctly proportioned fly. Houchmandzadeh *et al.* (2002) studied Bcd and Hb patterns in embryos grown at four temperatures from 9 °C to 29 °C. The Bcd patterns changed markedly with temperature; the Hb patterns did not. The nature of the change for Bcd was that the exponential decay constant k in $C = C_0 \exp(- ks)$ increased by a factor of about 2 from 9 °C to 25 °C (from my very rough measurements on the graphs). This corresponds to an apparent activation energy $E_a = 30$ kJ mol^{-1} for the constant $k = (k_d/D)^{1/2}$. The diffusivity D is likely to have $(E_a)_D \approx 12$ kJ mol^{-1}, whence $(E_a)_{k_d} \approx 72$ kJ mol^{-1}. With similar numerical work for my *Acetabularia* whorl spacings (Harrison *et al.* 1981) I got $E_k \approx 56$ kJ mol^{-1} for the rate constant k in the reaction–diffusion wavelength. However diverse the chemistries that these various rate constants refer to may be, one is likely to come up with E_as in the same ball-park if one is dealing with events having similar time-scales. Houchmandzadeh et al. (2002) concluded that variations due to temperature changes are compensated at the level of hb expression. They could perhaps have added that hb does not introduce any new temperature dependence of its own.

Lucchetta et al. (2005, 2008) used a technique called 'microfluidics' to maintain the two halves (anterior and posterior) of a Drosophila egg at two different temperatures (most usually 20 °C and 27 °C) throughout development, and looked at the progression of Eve (and Hb) pattern. All of the development was slower at the lower temperature, so the Eve stripes appeared first at the high temperature end. (They point out that when the posterior end was cool, the Eve sequence looks rather like that which the same group have observed in beetles.) But as development continued, all the Eve stripes formed and were in precisely the usual positions. This group also refers to a 'compensatory mechanism'. This mild phrase used both here and in Houchmandzadeh et al. (2002) somewhat downplays a big challenge to devise and confirm a mechanism to produce this absence of temperature sensitivity in the crucial patterning steps for segmentation. In regard to their approach, Lucchetta et al. (2005) conclude their account with a paragraph containing the following:

> Understanding the dynamics of a complex system by perturbing its environment in space and time does not require a priori knowledge of the system's components ... Perturbing the environment is a complementary approach to perturbing the molecular components of the network, as it might provide information on where and when events occur, rather than which molecules are involved.

I give enthusiastic applause to this statement, which expresses most cogently and succinctly a philosophy quite concordant with what I have tried to express in Section 1.5, including Fig. 1.1.

9

Various vertebrate events

What parts of the development of vertebrates are likely to have important relationships to the development of plants or of lower animals, and what parts seem to be so intrinsically different that no parallels need be drawn? I often hear talks in which the speaker refers in apparent generality to 'cells' or 'tissues' and I keep wanting to add 'animal' as a preceding quasi-adjective, because what is being presented does not apply to my preoccupation with plants. First, as already mentioned, the big contrast for many events is that in much of animal development, vertebrate and invertebrate, cells move past each other to change their relative individual positions, and also often move in orderly streams to position an aggregate within the organism. The latter clearly needs consideration of mechanical forces. But when cells move individually, these movements can be directionally random (or 'passive' in biological terminology) and must then obey a diffusion equation. In that case, the dynamics can become a larger spatial scale analogue of reaction–diffusion, one of the things I earlier referred to as 'cell-as-molecule' mechanisms (Harrison 1993). In Section 7.2.1 I described Malcolm Steinberg's work on cell sorting, illustrating such passive movement, but also a patterning that involves approach to thermodynamic equilibrium rather than departure from it.

Most animals develop the three-layer structure of endoderm, mesoderm and ectoderm very early in embryogenesis. In plants, I have mentioned interactions of two layers only as a possible mechanical effect, the set of elastic springs described in Section 4.2. While chemical interactions are also possible in plants, they are very well known in animals, in the form of 'induction', in which one layer affects the differentiation of cells in an adjacent layer. An instance of this is described in Section 9.1.

Amphibia are vertebrates that seem to have a lot going for them as subjects for developmental studies, especially in contrast to mammals. For example, they have huge eggs ($\times$10 linearly, $\times$1000 in volume, compared to mammals) that develop readily and visibly in natural fresh water; and limbs will regenerate after amputation. Considering such advantages, they are not studied as widely, in terms of number of species, as might be expected. The reason is that most species are quite difficult to keep going from generation to generation in the laboratory. Two exceptions are the most widely studied amphibian, the African clawed frog *Xenopus laevis* and, though by no means so commonly, a salamander, the axolotl (Spanish ajolote) *Ambystoma mexicanum*. It owes its Nahuatl name to its very localized natural habitat in the system of lakes that surrounded the Aztec capital (where it was used as food) and that have now been largely overwhelmed by Mexico City. Laboratory populations are all descended from a single set of specimens taken to France by a nineteenth-century French military expedition. One might have expected that this captive line would be all that now remains of this species; but there have been reports of its recovery during conservation efforts in the only remaining one of the Mexican lakes.

Amphibian heart development begins with classical induction of a region of the mesoderm from the underlying pharyngeal endoderm, causing mesoderm cells to differentiate in a manner described as 'specified' for myocardial cells. This means that they are making all the proteins needed to assemble the 'sarcomeric myofibrils' of a contractile system. The heart forms first as a simple tube from induced cells of the mesoderm. Fig. 9.1 shows cross-sections of the embryo of *A. mexicanum* at three stages, conventionally numbered 20, 24 and 28. Through these stages, the embryo is elongating and shrinking in diameter. The flank mesodermal regions shown on either side remain 1 mm long in the circumferential direction (and 1 mm long in the antero-posterior direction as well, so they are 1 mm square). Hence, as the embryo narrows, these flank mesoderms slide around as shown in the sequence of stages, until they meet at the ventral midline at stage 29, which occurs just after the stage shown in Fig. 9.1C. By that time, the region 'specified' for heart formation, shown cross-hatched, extends nearly halfway around the body, a region at least twice as wide as the part of it that actually goes on to make the heart. But if any part of that specified region is microsurgically excised and kept alive in a culture medium, it will form

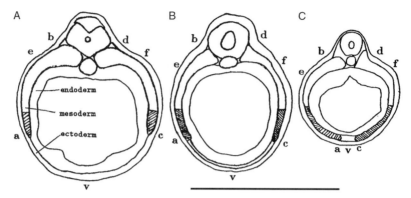

Figure 9.1 A cross-section through the embryo of the axolotl *A. mexicanum* at the level of the heart primordium at: A stage 20; B stage 24; and C stage 28. The bar is 1 mm, equal to the circumferential extent of the mesoderm on each side in all three sections. The shaded parts of the mesoderm are the regions specified for heart formation, i.e. they will form myocardial tissue if excised and cultured. Mesoderm edges a and c meet at the ventral midline v at stage 29.

a blob of tissue beating like a heart (Easton *et al.* 1994). The same sort of disparity between the specified region and what eventually makes the heart was observed earlier for other amphibia: the newt *Taricha torosa* (Jacobson 1960) and *Xenopus laevis* (Sater and Jacobson 1990). How, then, in the normally developing embryo, is heart formation inhibited over about half of the specified region? In *T. torosa*, there was some evidence for inhibition from the neural plate and fold, early stages of develop-ment of the nervous system on the dorsal side. But in the axolotl, heart formation starts long after neurulation. Induction takes three days, and another two days elapse before heartbeat begins.

John Armstrong, running a laboratory devoted to *A. mexicanum* at the University of Ottawa, suspected post-inductive dynamics within the specified region of the mesoderm, working to localize this organogen-esis. This is precisely the kind of event that Gierer and Meinhardt (1972) devised their formulation of reaction–diffusion theory to explain (Fig. 6.10). Fig. 9.2 shows this 'firing of a gradient' with the activator peak in the ventral midline and the gradient running laterally both left and right. In this kind of mechanism, of course, the inhibitor peak is coincident with the activator peak. I worked with Armstrong on an experiment–theory collaboration based on this concept, and the outcome was a large part of the PhD work of Holloway (1995; Holloway *et al.* 1994). It did not require all the horrors of 3-D computation and

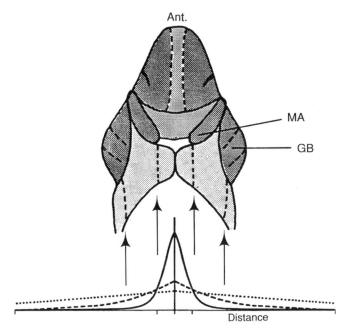

Figure 9.2 Sketch of the ventral side of the head of an axolotl embryo at stage 29. MA: mandibular arches; GB: gill bulges. The lightly shaded region is the flank mesoderm. On this, the outer pair of dashed lines (arrows) are the limits of the region specified for heart formation and the inner pair are the limits of the region determined for heart formation, i.e. the region where it will actually form in vivo. The concentration–distance graph shows the patterns that would be produced in each side of the mesoderm by a Gierer–Meinhardt mechanism (see Fig. 6.10). Activator A: solid curve. Inhibitor I: dashed curve. Source gradient: dotted curve.

simultaneous growth of the patterning region towards which he was led for his later work (Chapter 5).

Armstrong was keen that we should devise a mechanism with putative substances, but embodying general features of standard enzyme kinetics. What we came up with was the idea that endoderm-to-mesoderm induction produces an enzyme E in the mesoderm that catalyses conversion of a substrate S into an activator A, but that E requires allosteric activation by attachment of two molecules of A and one molecule of another substance R, to make the active complex EA_2R; an inhibitor I must also be produced to complete the Gierer–Meinhardt type dynamics. The activation of E by A_2 for production of the same substance A is the most obvious way to explain how the non-linear dynamics needed for reaction–diffusion patterning may arise.

The substance R (for 'rescuer') is something specific to the axolotl. There is a mutant (reported by Humphrey 1972) called cardiac-lethal, abbreviated as c. In mutant embryos (homozygous, c/c; the mutant is recessive) the specified mesodermal cells have all the proteins necessary for the contractile machinery, but those sarcomeric myofibrils never assemble. Contact between mutant and wild-type tissue – achieved by transplants between embryos or explants into culture medium – will, however, rescue the mutant tissue to make beating sarcomeres. This suggests a diffusible substance R in wild-type mesoderm, without which enzyme E is not activated.

The full putative reaction mechanism is (Holloway *et al.* 1994):

Equilibrium or rate constant

$E + I \rightleftharpoons EI$	K_1	(9.1a)
$E + 2A \rightleftharpoons EA_2$	K_2	(9.1b)
$EA_2 + R \rightleftharpoons EA_2R$	K_3	(9.1c)
$S + EA_2R \rightarrow EA_2R + A$	k_4	(9.1d)
$2A + S' \rightarrow I + 2A$	c'	(9.1e)
$S \rightarrow A$	a	(9.1f)
$I \rightarrow$ destruction	v	(9.1g)
$A \rightarrow$ destruction	μ	(9.1h)

The rate equations for this mechanistic scheme are as follows:

$$\partial A/\partial t = aS + kSERA^2/(1 + K_1I) - \mu A + D_A \partial^2 A/\partial s^2 \qquad (9.2a)$$

$$\partial I/\partial t = c'S'A^2 - vI + D_I \partial^2 I/\partial s^2 \qquad (9.2b)$$

D_A and D_I are diffusivities of activator and inhibitor, and s is distance along the system, in our case circumferential distance along the curve of the mesoderm.

These equations correspond quite closely to the Gierer–Meinhardt dynamics (Gierer and Meinhardt 1972; Meinhardt 1982, 1995). Meinhardt usually starts his solution to any problem by writing the rate equations, like Eq. (9.2a) and (9.2b), somewhat leaving the details of the putative chemistry to which they refer for readers to work out for themselves. The above may serve therefore not only as what Holloway, Armstrong and I cooked up for our specific problem, but also as a more general example of how rate

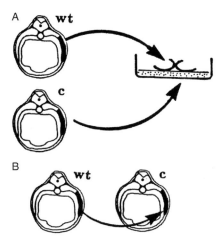

Figure 9.3 Schematic of explant and transplant experiments with wild-type (wt) embryos and cardiac-lethal (c) mutants, the latter being homozygous for the mutant gene, i.e. c/c, and both embryos being at stage 20. A: explantation of similar pieces of mesoderm from wt and c embryos and juxtaposition of the pieces in culture. B: unilateral transplant of wt mesoderm into a c host.

and chemical equations might be related in the Gierer–Meinhardt dynamics. (I write 'might' because, just as for Turing models in general, there is no unique mechanism to give the dynamics.) The general form that Gierer and Meinhardt have used for the non-linear term in the rate of formation of A is A^2/I. This is not obtainable exactly from any chemical mechanism, because it gives an infinite rate of formation of A whenever inhibitor concentration I is zero. The above example, with the form $A^2/(1+K_1I)$, is what usually turns up when one writes a chemical mechanism with an inhibition in it. Its dynamic behaviour will, in practice, be indistinguishable from that of the Gierer–Meinhardt equations except at very low I.

When one has devised such a mechanism, what kind of experimental evidence does one try to match with it? Much of the available data (Smith and Armstrong 1990, 1991, 1993) were from explant and transplant experiments, two of which are illustrated schematically in Fig. 9.3. In such experiments, formation of beating heart tissue can take variable time periods, sometimes different in the two pieces of mesoderm being observed, on the two sides of a living embryo or as two blobs of explant in contact in culture medium. There is also a more stochastic aspect to such results: some embryos form beating tissue, others do not. Both aspects were rolled together by Jacobson and Duncan (1968) into a

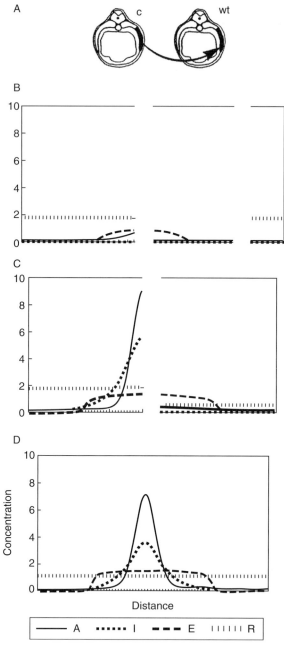

Figure 9.4 Results of a computation with a Gierer–Meinhardt-like mechanism, Holloway *et al.* (1994), matching the results of an embryonic stage-20 transplant rescue experiment, Smith and Armstrong (1993). A: one side (right, as shown) of the flank mesoderm of a c mutant embryo

single numerical variable, which was used by Smith and Armstrong (1990). This was the heart determination coefficient (HDC):

HDC = (% cultures that begin to beat) / (number of days to reach this state).

We found that we could not use HDC data. Our model would give an account of relative rates, so we could address the matter of 'number of days', but we concluded that percentage success in forming beating tissue could depend on a host of factors that were not in our model, e.g.: for in vivo experiments, variable rates of sliding of mesoderm over endoderm, and success in making contact at the ventral midline; for in vitro experiments, variability in tissue geometry and, again, success in making contact. Therefore, Armstrong had to supply Holloway and I with raw unpublished data on numbers of days that had been part of the published HDCs.

This illustrates an important general feature of collaborations between biologists and physicochemical people on projects involving dynamic theorizing. For the advancement of the interaction of experiment and dynamic theory, the most useful data are commonly quantitative changes in rates, times, sizes or number of parts in a pattern, rather than spectacular qualitative now-this-happens, now-it-doesn't kinds of observations. The latter have of course led via multitudinous mutations to vast amounts of information on genetics; but to address how the microscopic and macroscopic scales communicate in development, I believe that the physical chemist's propensity for getting really excited about quantitative changes is going to be necessary.

Fig. 9.4 shows a typical result of a computation of the dynamics of Eq. (9.2a) and (9.2b), corresponding to one of the experiments of Smith and Armstrong (1993). (Right and left in the following are from the reader's perspective.) A: at stage 20, a piece of mesoderm has been transplanted from the right side of a cardiac-lethal (c) mutant embryo to the same location in a wild-type (wt) embryo. A short patch of wt mesoderm remains on the right, dorsally; this is extreme right on the

Caption for Figure 9.4 (cont.)
is transplanted into a wt embryo. B, C and D: from the computation for the mechanism of Eq. (9.1) and (9.2), concentration–distance plots at stages 20, 29 and 35 (when the heartbeat is observed experimentally), respectively. The distance scale around the mesoderm is centred on the ventral midline, with gaps left where parts of the mesoderm are not yet in contact, i.e. centrally before stage 29, and towards the right in B, just after the transplant. See text for explanation of dynamics.

graphs (B, C, D). B: at stage 20, both sides have a small part of the mesoderm induced (broken lines); the c mutation does not hinder induction. But growth of an activator (A) peak (solid line) has started only on the left, not in the transplanted mutant piece. This has no rescuer (R); but the remaining bit of wt mesoderm on the right has R. C: at stage 29, when left and right mesoderms are about to touch in the middle, induction has given E to a wider region (too wide for the size of the heart); R, which is assumed in this model to be very fast-diffusing, has spread quickly over the right side to give a low concentration and hence very slow beginning of the growth of an A peak. D: stage 35, R is present at high concentration everywhere, having diffused from left to right across the ventral midline when the two parts of the mesoderm touched at stage 29. Growth of the A and I peaks has now caught up, to give an A peak covering the correct size of the region for heart formation, as in Fig. 9.2.

For details of results of a variety of transplant and explant experiments, see Holloway et al. (1994). For times to beat (giving our computations a time-scale by assuming that they gave the right time of 2.1 days for wt development) we got good matches in other types of experiment for times from 2.1 up to 9.6 days.

In Chapter 6 I described the Wigglesworth (1940) 'inhibitory field' concept, and the need to elaborate it into Turing dynamics if the activator also diffuses. The above account is an example of that progression. Sater and Jacobson (1990) suggested the Wigglesworth concept for localization of the heart-forming region by action from the ventral midline rather than the dorsally located neural structures. Armstrong (1989) suggested a Turing model, to address, for example, rescues; and got Holloway and I involved in elaborating it. To me, what matters here is that the step from Wigglesworth to Turing should be regarded as a very normal little step that happens often and easily in such work, not a jump needing the attention of the great mythical heroes.

9.2 THE ANGEL IN THE FISHBOWL, AND OTHER SETS OF STRIPES OR SPOTS

Should one study what, in embryological development, unites the horse and the zebra, or what superficial process eventually makes the spectacularly obvious difference between them? I have chosen to take the bull-at-a-gate direct methods, seeking to apply kinetic theory to crucial events of embryogenesis (and of organogenesis throughout life in plants). But if there are unities to be found in developmental mechanisms of pattern formation, they may just as readily be revealed in the first instance by studying superficial decorative features that are easily observed. Such work has led to substantial advances in dynamic theory. Murray (1981a, 1981b) applied to animal coat patterns a pair of reaction–diffusion equations that seem to have a particular sensitivity to the shape of a 2-D region, choosing to make spots on a squarish region and stripes across a long narrow one. From this work, he claimed in a footnote the status of a theorem for the statement: 'It is not possible to have a striped animal with a spotted tail; the converse is quite common.' Perhaps someone should study the efficacy of humour in the devising of mathematical theorems.

Stripes-versus-spots is, however, a very substantial problem capable of commanding the attention of many theorists. One significant aspect is diversity in patterning behaviour for diverse kinds of non-linearity in reaction–diffusion mechanisms. The proper approach to this requires one to plunge into the equations, because one has to look at the morphogen concentration variables differently. When Turing (1952) devised his reaction–diffusion equations, he was considering the problem of how spatially repeating patterns can arise from an initial spatially uniform (often called homogeneous) steady-state. Let us call the morphogen concentrations in that state X_0 and Y_0. The first stirrings of non-uniform pattern are small departures from that state; so let us discuss the matter in terms of those departures, $U = X - X_0$ and $V = Y - Y_0$. If the rate equations for any particular reaction–diffusion model are written first as equations in X and Y, and then as equations in U and V, the non-linearities in the equations will look substantially different in the two sets of equations. For instance, the Brusselator X, Y equations contain only the non-linearity X^2Y, a cubic term; but the Brusselator U, V equations contain also quadratic terms, UV and U^2. It has been known for some time that, in 2-D patterning, mechanisms that give quadratic U, V terms are 'spotting', while those that give only cubic U, V terms are 'striping'. This was suspected on a computational basis by Nagorcka (1988), Lacalli (1990) and Lyons et al. (1990). Meanwhile, two analytical pieces of work were being done simultaneously and quite independently, by different methods, by Lyons in my group and Ermentrout, leading to simultaneous publications (Lyons and Harrison 1991; Ermentrout 1991).

The significant feature of the kinds of non-linearities that have a strong tendency to make stripes can be expressed as a very simple symmetry property (Fig. 9.5) (Lyons and Harrison (1992a) being an account intended for biologists – except the Appendix!) If one looks at

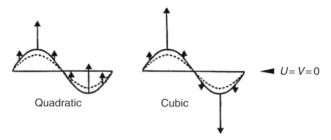

Figure 9.5 Symmetries of pattern growth related to 'stripes versus spots' in 2-D patterning. In rate equations converted from X, Y to U, V terminology (see text), quadratic terms and cubic terms give progressions from initially sinusoidal pattern as shown by the arrows, and correspond respectively to spotting and striping tendencies.

the early sine waveform of emerging pattern, in which crest and trough half-waves are of the same shape, mechanisms that are striping (cubic terms only) have equal growth rates upwards for the crests and downwards for the troughs, because all the terms in the U, V equations are odd for sign-reversals of U and V together. Such dynamics I referred to as having a kinetic chirality (sections 8.2.3 and 9.2 of Harrison 1993), and a mechanism of my own devising, which I called 'hyperchirality' has that property (Harrison and Lacalli 1978).

Lyons went on (Lyons and Harrison 1992a, Appendix, and 1992b) to consider another kind of mechanism having a similar symmetry property, but not a reaction–diffusion mechanism. This work addressed a particular problem of patterning in the brains of primates and some other vertebrates. In the primates, the optic nerves from each eye make equal numbers of connections to the right and left sides of the primary visual cerebral cortex (on the back surface of the brain). These connections are patterned as 'ocular dominance columns', which are effectively a pattern of stripes each about 1 mm wide, all extended into a third dimension without further patterning. How is this pattern generated? A general concept is that synapses connecting message-carrying neurons, identifiable as carrying messages from the left or right eye to cortical neurons, can interact laterally to strengthen or reinforce each other. A model of Swindale (1980) postulates a symmetry for these interactions (Fig. 9.6): the interaction between like (same eye, ipsilateral) synapses is positive at short distances and negative at longer distances. The interaction between unlike (opposite eye, contralateral) synapses is just the opposite. These interactions have 'kinetically chiral' symmetry. For this model, just as he had for reaction–diffusion models

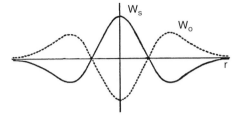

Figure 9.6 In the eye–brain connection, symmetries of interactions between two synapses as postulated in the model of Swindale (1980). Abscissae r are distances between the synapses (in a single layer of the primary visual cortex). Vertically, upwards is favourable interaction, downwards is unfavourable. W_s, solid curve: interactions between synapses for optic neurons both from the same eye ('ipsilateral'); W_o, broken curve: interactions between synapses from opposite eyes ('contralateral').

lacking quadratic terms, Lyons analysed systems of solutions for sets of stripes all of one wavelength but with different orientations, as to whether the stability condition was for one set to dominate (striped pattern) or, say, two sets at right-angles or more complex combinations of any number of orientations to develop equally (spots). In both cases, reaction–diffusion and the ocular dominance model, he found that the stable pattern was a single set of stripes. But also, in relation to the general philosophy of kinetic theory of patterning, the striking part of his analysis was that in the two cases the stability equations were not just similar, they were identical. There can be no better illustration than Lyons' progression in this piece of work of my contention that study of reaction–diffusion is the one well-paved highroad into kinetic theory, and is the way to go even if, in the end, you never find an activator–inhibitor pair of molecules in the simplicity of Turing's original idea.

The need for kinetically chiral symmetry to make stripes is well illustrated by the contrast between ocular dominance patterns in a primate, the macaque monkey (Fig. 9.7A), in which the mix of ipsilateral and contralateral connections is 50/50, and the cat (Fig. 9.7B), in which the mix is 70/30 in favour of contralateral, and the resulting pattern is closer to spotted than striped.

The cubic-versus-quadratic rule for stripes-versus-spots is not inviolable, and is not the whole story. Borckmans et al. (1995), in comprehensive studies of the Brusselator in 2-D domains, obtained instances of both stripes and spots. As shown in Fig. 8.5, Lacalli forced a Brusselator to make stripes by putting gradients of rate constants

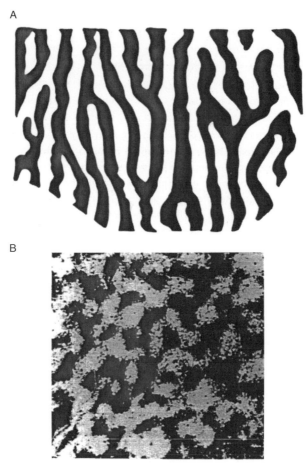

Figure 9.7 Patterns of ocular dominance stripes or patches: A, macaque
monkey; B, cat.

along the domain. This worked for wider domains than that shown
(e.g. Fig. 3B of Lacalli *et al.* 1988).

A non-living chemical system that produces either striped or
spotted pattern in different regions of concentrations by a fairly well-
established reaction–diffusion mechanism is the CIMA (chlorite–
iodide–malonate) reaction in a gel reactor (Lengyel and Epstein 1991).
Stripes were observed at high initial iodide or low initial malonate
concentration (Ouyang and Swinney 1991). I found that, for the published mechanism, these conditions corresponded to a high ratio of
cubic-to-quadratic terms in the U, V equations (published general discussion, pp. 412–13, accompanying Harrison *et al.* 2001).

I have long maintained that the principal evidence that patterning is happening as a result of dynamics must come from the dynamic behaviour of the patterning phenomenon itself, not from a knowledge of the molecular species involved in the process. This is, I think, an attitude having the flavour of a physicist's philosophy, because it corresponds to what was going on in great advances in the study of light between the seventeenth and early twentieth centuries, throughout which period the particulate nature of light was not recognized. The applicability of Newton's laws of motion to bodies great and small of unknown compositions but known masses is another example.

That patterning dynamics can reveal patterning mechanism seems to have been accepted by whoever wrote the cover headline on an issue of *Nature*, 'Turing patterns come to life', calling attention to Kondo and Asai (1995), 'A reaction–diffusion wave on the skin of the marine angelfish *Pomacanthus*'. The phenomenon is that as angelfish – of two species in this genus – grow, the number of stripes in their skin pattern increases in such a way as to keep the spacing between stripes constant. In detail, various ways are observed in which the patterns go about their readjustment of spacing. In the species *Pomacanthus semicirculatus*, three stripes were the pattern as the fish grew from 2 cm to 4 cm long, so that the spacing between the stripes doubled. At that point, extra stripes were inserted so the original spacing was restored. This is akin to what Holloway did in his 2-D computations of simultaneous growth and patterning in *Micrasterias*: he made the program insert a new node when any internode had doubled in length. But in the angelfish, the patterning process is writing its own program to do this – the Organizer again! In *P. imperator* (Fig. 9.8), a single new stripe was usually inserted into the pattern of parallel stripes by formation and movement of a defect in the pattern. Fig. 9.8A (detail I), B–D (the fish) and E–G (computation) show what I call a 'tuning fork' defect travelling from posterior to anterior, replacing one stripe by two. Kondo and Asai point out that this requires readjustment of the positions of all the stripes to restore the original uniform spacing when four stripes have been replaced by five. In the real fish, the process took more than three months, and in the computation 50 000 iterations. Fig. 9.8A (detail II), H–L (the fish) and M–Q (computation) show two branching points from four to five stripes meeting and annihilating each other, another equally slow process. The computations were done with a simple linearized Turing model.

It is evident that, for these particular species of fish, morphogenesis measures rather than counts. The patterning mechanism, whatever

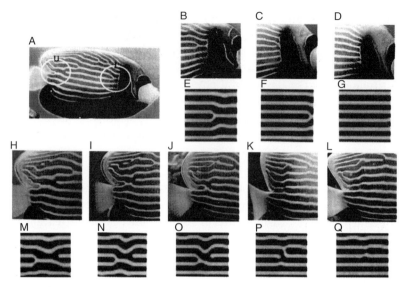

Figure 9.8 Rearrangement of the stripe pattern of the angelfish
Pomacanthus imperator. From Kondo and Asai (1995) with permission. A: an
adult, ∼10 months old. B–D: close-ups of region I of A, at the same time and
two and three months later. E–G: reaction–diffusion computer simulation
for region I, starting pattern, after 30 000 iterations, after 50 000 iterations.
H–L: close-ups of region II in A, at the same time and after 30, 50, 75 and
90 days. M–Q: computer simulation for region II, starting pattern, and result
after 20 000, 30 000, 40 000 and 50 000 iterations.

it may be, can determine a spatial scale of distance, as can *Acetabularia*
in forming whorls of vegetative hairs, as can *Larix* x *leptoeuropaea* in
forming cotyledons; but unlike *Drosophila* pair-rule stripes, which are
counted to precisely seven, against 30% variation in embryo length.[1]

9.3 SOMITES IN SPACE AND SEQUENCE

Here I return to the topic of unity versus diversity in segmentation mechanisms, which I addressed from the viewpoint of arthropods
in Chapter 8. The generalities about segmented somite formation in the
vertebrates are: neurulation proceeds by forming the neural tube along
the dorsal midline, with paraxial mesoderm on either side of the neural
tube. This eventually resolves into three dozen or more (according to

---

[1] As discussed in Chapter 8, reaction–diffusion can count if there is an underlying
prepattern.

species) separate and clearly bounded groups of cells called somites, on each side, and precisely paired on the two sides (Fig. 9.9). They are transient structures, but are crucial to the segmental organization of the vertebrate body, which exists in the nervous system and musculature, but in later life is obviously segmental only in the vertebrae. There is both a spatial and a temporal aspect to the problem of somite formation mechanism. Somites form sequentially, from anterior to posterior, at regular intervals: 6 h per somite pair in the human embryo, for the formation of 38 pairs of somites from days 21–31 of development; 90 min per somite pair in the chick. In most species, at any given time during somitogenesis one can usually see a group of about six somites on each side at various stages of formation, apparently developing sequentially from anterior to posterior. The antero-posterior length of a somite is usually much the same for all pairs, leading to my usual question: what's doing the measuring of length, or counting of cells, for one somite?

In the development of theories of somite formation, the uniform time sequence of formation has been the starting-point and has led to the concept that there is a clock mechanism involved. The molecular components that seem to be essential for the clock are homologous to *Drosophila* genes and their products: the Notch protein and its receptor, which take part in cell-to-cell signalling, and *hairy*, a *Drosophila* pair-rule gene. For a review mentioning vast numbers of substances possibly involved, as well as the reaction–kinetic models, see Rida *et al.* (2004). It is a highly condensed account, giving guideposts into a vast literature. For a physicochemical account with equations, describing a model that includes the segmental clock but deals chiefly with reaction–diffusion–advection, that last word referring to a flow from an upstream (anterior) boundary, see Kaern *et al.* (2000, 2001) and Jaeger and Goodwin (2001).[2]

Since, in relation to arthropod segmentation, Peel (2004) mentioned the possibility that a Notch-dependent segmental clock perhaps existed in earlier short germ band insects but was lost in the Diptera, I, with my immutable way of thinking, was led to playing with the idea that a somite formation mechanism might be able to work without a cellular clock. The group of about half a dozen somites in various stages of formation suggested to me a correlation with Lacalli's Brusselator-with-a-gradient computation shown in Fig. 8.5. Is there a simple way to make this group move progressively posteriorly as the somites at the

[2] Also see Palmeirim *et al.* (1997); Pourquie (2003); Baker *et al.* (2009).

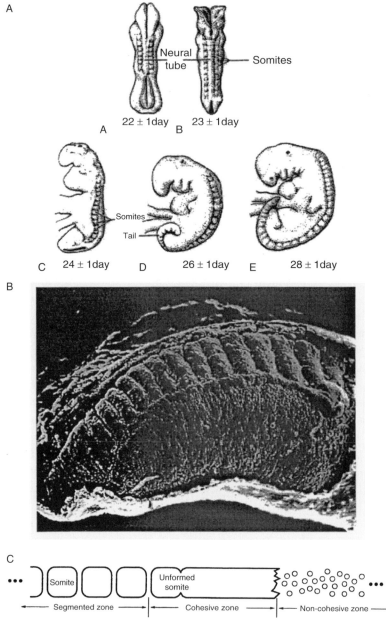

Figure 9.9 Formation of somites. A: developmental sequence in a human embryo from 22 to 28 days, showing pairs of somites forming sequentially from anterior to posterior. From Moore (1974) with permission.

B: scanning electron micrograph of an axolotl embryo at stage 29 showing the right-side row of somites (i.e. anterior is right in this picture). Compare anterior end of the group reach complete formation? For this, I assumed that a somite achieves complete formation when a Brusselator activator peak X reaches some threshold value (Fig. 9.10, lightly dotted line) and that the peak then stops growing and the new somite starts producing the reactant B (or the enzyme for rate constant b). This diffuses posteriorly and decays, to make an exponential gradient, like Bcd in *Drosophila* and doing the job of Lacalli's gradients in the computation for Fig. 8.5. Fig. 9.10 shows the progression of the pattern at four times: 0, 45, 48.6, and 192.75. The short gap from 45 to 48.6 is to show the first somite peak just before and after maturation, with the shift of the head of the B gradient from the left-hand edge to end of the first somite. Fig. 9.10 is from the work of Jefferey J. Orchard, to whom I gave this problem as part of an MSc project. It has previously been published only as an abstract (Harrison *et al.* 1996) and in the general discussion, pp. 345–6, accompanying Harrison *et al.* (2001). The discussion includes Kaern's reply to my question as to whether this model captures much of the dynamic nature of somite formation. He said that it could not account for observed gene expression waves in the chick, mouse and zebrafish, but that a segmental clock has not yet been identified in reptiles. Also, he believed that it did not account for observations that the direction of somite formation can be reversed by inverting a piece of the presomitic mesoderm.

This work was to have been the start of my next collaboration with Armstrong's group after the heart formation project described in Section 9.1, and I had some preliminary information to start modelling. This was that in the axolotl, the somites are formed in a 'cohesive zone' of the presomitic mesoderm, in which the cells are joined by gap junctions and capable of communicating diffusively (Armstrong and

Caption for Figure 9.9 (cont.)

Figs. 9.1 and 9.2 for other perspectives of the axolotl embryo at similar stages. From Armstrong and Malacinski (1989) with permission.

C: schematic of the three zones present along the paraxial mesoderm in the axolotl during somite formation. Anterior is left in this picture, and the boundaries between the zones move to the right as somites form sequentially. Segmented zone: somites are completely formed. Cohesive zone: cells cohere and can communicate cell-to-cell via gap junctions. About five somites are at the intermediate stages of formation. Non-cohesive zone: cells are essentially separate and somite formation has not started. (Information from Armstrong and Graveson, 1988, and personal communication with Armstrong.)

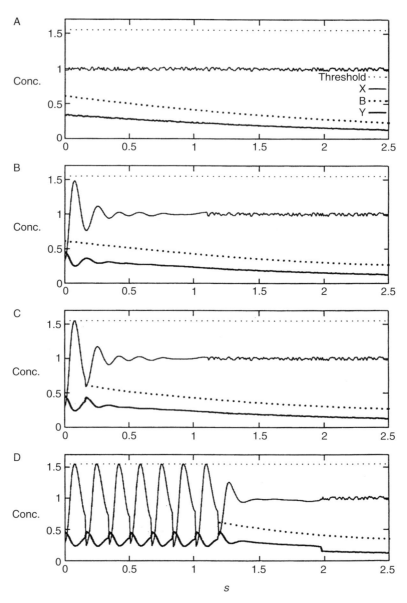

Figure 9.10 Computational results for production of somites sequentially, but with about five at a time in the 'cohesive zone', based on the Lacalli computation for a Brusselator with a gradient as shown in Fig. 8.5. Here, the gradient is in the concentration of the reactant input B (dotted line, heavy dots). It is a local source – diffusion – first-order decay gradient (steady-state exponential, as in discussion of Bcd in Chapter 8); but the local source is the posterior edge of the last-formed somite, so it moves

Graveson 1988). Posterior to the cohesive zone there is a non-cohesive zone in which the cells are migratory. My use of B for the exponential gradient, instead of Lacalli's c, was because B can control whether the dynamics are in a pattern-forming or unpatterning region of the Turing parameter space. I identified this boundary with the end of the cohesive zone (beginning of noisy part of the X profiles in Fig. 9.10). Most unfortunately, ill-health forced Armstrong to close down his axolotl laboratory, and no long-term collaboration arose.

I present this rudimentary model, despite its apparent deficiencies, because I think it is useful in exploring complex dynamics to apply Ockham's razor and find out what is the simplest model that has the capacity to do the job. This may help in the quest for the unity in segmentation processes.

9.4 GASTRULATION: GEOMETRY, TOPOLOGY OR CYTOLOGY?

Unlike the majority of specific topics I have chosen to address in this book, gastrulation is not a process upon which I have been involved in any experimental or theoretical work myself. Nothing that I say about the potential of dynamic theories for its explanation uses my own work as a take-off platform. But the event is crucial to the establishment of the body plans of most metazoa (multicellular animals that have embryonic stages of development), so I could hardly ignore it. When I first moved into the developmental field in the early 1970s, I was soon struck by the fact that, when some zoologists said 'morphogenesis', they were often using the term almost synonymously with 'gastrulation', regarding that phenomenon as presenting the major

Caption for Figure 9.10 (cont.)

antero-posteriorly as each new somite forms. A somite is defined by an X-morphogen wave, centred on the peak, and is fully formed when the peak reaches a threshold height, here 1.55 (dotted line, light dots). The other end of the cohesive zone is defined by a threshold value of B below which cells do not have the cohesive property (i.e. ability to form gap junctions). In this computation, this value is $B = 0.4$. Except at the start of the computation (A), noise is added to X and Y only in the non-cohesive region. This model does not contain a clock explicitly, but the formation of each somite when X reaches its threshold value is implicitly a clock, with no mechanism given. Previously published on p. 346 of general discussion accompanying Harrison *et al.* (2001).

challenge for the explanation of animal development. (It did not help my state of confusion with biological terminology that my closest associate in my change of field, Thurston Lacalli, though registered in zoology, had just completed a PhD thesis on a plant, 'Morphogenesis in *Micrasterias*', which gave me quite a different slant on the meaning of the word.)

The word 'gastrulation' implies the formation of a stomach, and by implication all the rest of a digestive system, a major feature of material intake from the environment for most of the animal kingdom. The formation of this channel along the inside of the body occurs very early in development of both vertebrates and invertebrates. But is there a precise definition of gastrulation, such as might help in looking for some level of unity in its mechanism? On browsing text-books, I found that the essential nature of the event most commonly emerges from detailed descriptions as the formation of the three germ layers, ectoderm, mesoderm and endoderm, and that the formation of a channel right through the body is not necessarily mentioned at all. I had thought that the latter was primary: it is a great change in topological character of the shape of the body, from that of a sphere (simply connected surface) to that of a torus (non-simply connected surface). My instinct is to think first of big parts and how they are formed by subdivision of the whole organism. That can lead to a pre-occupation with precise details of geometry, as must have become evident to the reader from my obvious desire to find Bessel functions and spherical harmonics in plants.

But any cursory comparative look at the details of the gastrulation event in such popular beasts as nematodes, fruit flies, sea urchins, amphibians and chickens will quickly show that the geometrical changes are messy and immensely diverse. The unities are: topological, in the relentless progress from a ball to a doughnut; cytological, in the differentiation of three types of cell, ectodermal, mesodermal and endodermal, and further in the digestive function of a large fraction of the endodermal cells. To my favourite question, 'Does morphogenesis measure or count?', should perhaps be added another: 'Does morphogenesis observe (and respond to) geometry or topology?'. And also, as one contemplates some of the major detailed diversity, another version of the same question I asked in relation to segmentation processes in Chapter 8: 'At what level may there be any kind of unity in the mechanisms of gastrulation?'. As an example of the lowest level of multicellular organization at which such a unity might be found, Wolpert (1990, 1992) suggested differentiation of two cells from each other, one into a cell specialized for feeding, the other for division and hence reproduction.

In his 1992 article, he describes also the view of the late-nineteenth-century evolutionary biologist Haeckel on the origin of the metazoa. (Haeckel invented the word 'phylogeny' and is notorious for his addiction to drawing trees of life with knobbly branches complete with bark.) His view was that some protozoan became colonial in the form of a hollow sphere ('Blastea'), which developed into a two-layer structure (Gastrea) formed by an inward bulging (invagination) of part of the spherical shell, giving rise to the inner layer. Wolpert mentions a somewhat different view in which the inner layer forms by ingression of cells, i.e. a flow of disconnected cells that do not retain their positions in a layer as the flow occurs. In existing species, examples of both invagination and ingression are well-known. But gastrulation does not always start from a hollow sphere (which, when it does exist, is usually called a blastula). The sphere, as a precise geometrical figure, has been over-emphasized, not least from my several mentions of it in the previous paragraphs. For instance, Turing (1952) seems to have thought that spherical blastulae were a common feature of vertebrate development. He wrote: 'An embryo in its spherical blastula stage has spherical symmetry ... But a system which has spherical symmetry, and whose state is changing because of chemical reactions and diffusion, will remain spherically symmetrical for ever ... It certainly cannot result in an organism such as a horse, which is not spherically symmetrical.' And towards the end of his paper he discussed spherical harmonics in relation not to plants, as I have done in Chapter 4, but specifically in regard to gastrulation. In observable fact, anything really close to a single-cell-layer spherical blastula occurs only in the echinoderms, such as the widely studied sea urchins, and even then arises with sufficient difference (ectodermal versus endodermal) between its two halves that a symmetry-breaking event is no longer necessary. This is also the instance in which gastrulation occurs by a deep invagination forming a tubular archenteron, the most obvious progression from sphere to torus as soon as the archenteron connects with the opposite side of the spherical shell.

Fig. 9.11 is from Wolpert (1992) and attributed as 'after Tardent (1978)'. In modern textbooks, e.g. Gilbert (2006), gastrulation is usually lavishly illustrated and meticulously described phylum by phylum. I present this single figure, and it is not about a three-germ-layer organism, but an illustration of diverse ways of forming a two-layer structure, called the planula, within the phylum Cnidaria (or Coelenterata), which includes the well-known *Hydra*. In this phylum there is no mesoderm, but a jelly-like layer with very few cells in it called the

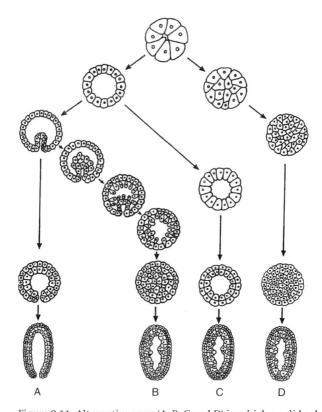

Figure 9.11 Alternative ways (A, B, C and D) in which a solid sphere of cells
can develop into a hollow elongated two-layer structure, as happens in the
phylum Cnidaria (Coelenterata), where this structure is called the planula.
This diagram effectively summarizes the alternative ways in which
gastrulation can occur in a wide range of invertebrates and vertebrates
that have a three-layer structure. From Wolpert (1992) with permission.

mesogloea, between ectoderm and endoderm. Nevertheless, the diversity of ways of advancing from a one-layer to a two-layer structure gives
a good impression of how confusing it can be to approach the literature
of gastrulation as a beginner, especially if one does so with some
pig-headed idea that a common mechanism is to be found somewhere.
Fig. 9.11 does not show the topological sphere-to-torus transition; this
happens if there is a breakthrough to form a gap at the top of the last
stage shown for Fig. 9.11A.

Both above and below the echinoderms, gastrulation displays very
different geometries. In some instances, there is a mass of cells that is
solid rather than hollow, in which a cavity (blastocoel) appears, but not

to the extent of thinning the mass to a single sheet. A very simple case is the nematode (roundworm) *Caenorhabditis elegans*, for which, as successive cell divisions occur, the destiny of every cell is always exactly the same. At the 24-cell stage, one cell is designated E because it is the sole precursor of intestinal endoderm. It is on the outside of a somewhat elongated mass of cells in which a small blastocoel appears only transiently. The cell E divides into two, Ea and Ep, which ingress to the location of the blastocoel and go on to form a 20-cell gut.

Amphibia, according to Gilbert (2006), have diverse ways of going about gastrulation; but if some generalization can be drawn, it is that there is a spherical structure that is one cell layer thick over part of its surface, but much thicker over the remainder; the thin part becomes ectoderm; the thick part becomes endoderm. Between these two parts is a cavity called the blastocoel. It is moved aside and replaced by a new cavity, the archenteron, as mesoderm moves in by a process that looks somewhat intermediate between invagination and ingression in many of the diagrams in textbooks.

Vertebrates mostly contrive to give a different meaning to the word sphere, i.e. the topological meaning rather than the geometrical. Gilbert (2006), having earlier stressed the species-specific creativity of amphibia in how they go about gastrulation, manages, in a single account of the process in the mammalia, to write of mice and men as if they were the same species. So perhaps I may be permitted to generalize: vertebrates tend to start gastrulation from a single more or less flat layer of ectoderm overlying a large mass of yolk. From one end of this layer, a second layer of endoderm is inserted between it and the yolk, eventually forming a rather flat bag, with the ectoderm as one side and the endoderm as the other. If one is a topologist, and uses the word 'sphere' to mean 'simply connected surface', so that a cube or a flattish bag is a sphere, then the vertebrate starting-point for gastrulation is no different from that of the amphibian or the echinoderm: a topological sphere, with the two sides differentiated from each other as ectoderm and endoderm. From this start, a complex method arises for mesoderm to ingress and fill the bag. This, as well as some of the details of what happens to the more geometrically spherical bag in the amphibia, has more or less dictated the scientific questions that much of the best experimental work on gastrulation has been designed to answer. In brief, these questions are largely about the qualitative roles of cell-to-cell interactions in differentiation, and the significance in this respect of some quite small groups of cells forming morphogenetically important structures, e.g. the Spemann organizer, the Nieuwkoop centre and

Hensen's node. The first two of these were discovered in amphibia; the node is a mammalian structure (and, most curiously, nowadays is called node without personal attribution in the mammalia, but carries the name of Hensen in birds; he first found it in rabbits.)

In the amphibian blastula – more or less geometrically spherical externally though partly empty and partly filled with endoderm cells inside – ingression of mesoderm cells takes place through a small opening called the blastopore. One side of this (the dorsal blastopore lip) is the Spemann organizer, discovered to have powerful developmental properties by Spemann and Mangold in 1924. By transplantation experiments on this bit of tissue between two species of newt, they were able to show that this organizer has a strong role in specifying the directionality of the antero-posterior axis along the dorsal side of the embryo, first seen morphologically by formation of the notochord and the neural tube. This influence of one group of cells upon another is referred to as primary embryonic induction. Later, Nieuwkoop (1969, 1973, 1977; see also Nieuwkoop *et al.* 1985) identified a group of endo-dermal cells that induce the organizer.

For birds and mammals, consider a flattish bag with a more or less rounded outline, the outer side of it is the ectoderm, the inner (i.e. nearer to the yolk mass) side is the endoderm. From the posterior end of it, a narrow, thickened region called the primitive streak develops anteriorly, ending (behind what will become the head) at a small region called Hensen's node, or just 'the node', which is the equivalent of the organizer or dorsal blastopore lip in amphibia. Ingression of cells to fill the bag, especially those that will become mesoderm, takes place largely through the node, but also through other parts of the streak. I have used words like 'anterior' and 'dorsal' as if these directions already existed; they may, but alternatively it may be the formation of the streak that defines what these directions are to be.

Most biological accounts of gastrulation are devoted to meticulous descriptions of detail in each type of organism addressed separately; my above abbreviation, making every case look like a variant of the same sequence, is likely to make the authors of these accounts shudder. I hope, however, that it may assist readers from the physical sciences if they are approaching this topic for the first time and trying to ingress themselves into the biological literature. One generalization that emerges from a survey of that literature is that the kinds of evidence that are most readily acquired lead towards recognition of: the roles of highly localized groups of cells; the listing of numerous kinds of induction from these to larger regions, from layer to layer, and so forth; the movements of cells; and the differentiations of cell types. ('Readily acquired' may be a great over-statement of the ease of obtaining such information; transplantation experiments on extremely tiny bits of tissue in embryos that are overall quite tiny are very difficult.) These kinds of evidence are essentially qualitative. To get spatially quantitative data that might help to determine whether there are dynamic pattern-formation mechanisms at work is extremely difficult, as well as not being the information that many experimentalists are most anxious to seek. It is in later stages of animal development that one finds that a region induced for a particular organogenesis is not equal in size to the region that makes the organ, as I illustrated for the axolotl heart in Section 9.1.

Meinhardt (2001), however, has published a mechanistic scheme showing step by step how a chain of activation–inhibition reaction–diffusion processes could account for polarization of the embryo, formation of the Nieuwkoop centre, formation of the three germ layers, generation of the Spemann organizer by interaction with the Nieuwkoop centre, conversion of the organizer into a notochord, and origin of a left–right disparity. There has also been an indication from the Stern group that, in the chick embryo, 'an interplay of inducers and inhibitors must be involved in determining the position of the primitive streak' (Streit *et al.* 2000). The Stern group went on to collaborate with Maini's theoretical group in devising a model of primitive streak initiation in the chick embryo that starts from reaction–diffusion but goes beyond the usual concept of a chemical pattern that reaches steady-state by having as one of its major features a travelling wave of inhibition (Page *et al.* 2001). This report includes suggestions for the possible chemical identities of activators and inhibitors, related to the 'Wnt signalling pathway'. (The name 'Wnt' is a combination of the initial of 'wingless', a *Drosophila* gene as usual named pejoratively for what happens when it is inactive, and a couple of letters from 'integrated', a vertebrate homologue of *wingless*.) I refrain from listing the several suggestions for inhibitors because I do not wish to emulate the unfortunate habit of many geneticists of throwing new and quite unfamiliar names at a general audience far faster than anyone can commit to memory new words in a foreign language.

Marée and Hogeweg (information from a talk by Marée)[3] are developing a model of chick gastrulation that specifically addresses

---

[3] See also Käfer *et al.* (2006).

the streaming movement of cells in the formation and continued action of the primitive streak using the Graner and Glazier method of tackling cell movements dependent on differential adhesion (Section 7.2.1, especially Fig. 7.3). This model also involves the fibroblast growth factors Fgf4 and Fgf8, produced respectively by node cells and streak cells, both decaying and both diffusing, with Fgf4 acting as a chemoattractant for streak extension. Modelling of this kind is beginning to put together motifs from reaction–diffusion and mechanochemistry.

9.4.1 A note on cellular slime moulds

Marée, in his PhD thesis work (Marée 2000; Marée and Hogeweg 2001, 2002) had first applied the Graner and Glazier method to streaming cell movement in a very different organism. The cellular slime mould *Dictyostelium discoideum*, a common organism in soil, exists as single amoeboid cells during the feeding part of its life cycle. When it has used up its food supply, the cells form aggregates that assume an elongated form that can crawl over a surface (therefore known to everyone as 'slug', despite earnest attempts to popularize 'pseudoplasmodium'), within which the cells differentiate into two types called 'prespore' and 'prestalk' cells. The slug stops moving and rounds up (Fig. 9.12A), and the cells then stream in a 'reverse fountain' (arrows in Fig. 9.12), driving a central downward movement of stalk cells, while the spores form a spherical mass on the upper part of the stalk. Here, in a different kingdom from the animals (and the citizenship of cellular slime moulds is somewhat uncertain), we have successive processes, aggregation, slug migration and culmination, that all demand theory of cell movement similar to what is needed for gastrulation.[4]

9.5 IS THAT ALL THERE IS?

Of course not. The diversity of organs that develop very differently in the vertebrate body is immense, and each is made by many patterning events. My aim in writing this book has nothing to do with comprehensive coverage, but to select a few events that may highlight the problem of whether kinetically generated and maintained pattern is involved, and especially what I believe to be the modes of thinking needed to approach this possibility. Yet, having decided to make an end at this chapter, I am left with a guilty conscience regarding the

---

[4] See also Palsson and Cox (1997) and Palsson (2007).

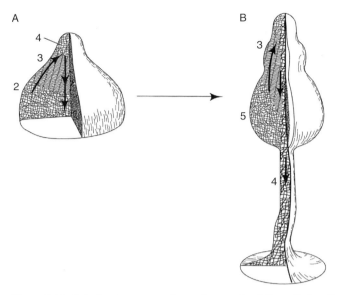

Figure 9.12 Culmination stage of morphogenesis of the cellular slime mould *Dictyostelium discoideum*, to illustrate concerted flow of streams of cells, here the 'reverse fountain' indicated by the arrows. Pre-spore cells (2) mature into spore cells (5); pre-stalk cells (3) mature into stalk cells (4), which generate a heavy cellulose sheath for mechanical stability of the stalk.

number of well-studied developmental events that I haven't so far mentioned at all. Here are some brief comments about a few.

9.5.1 Vertebrate limb development

The vertebrate limb (very commonly studied in the chicken) starts to develop as a bud, containing a mass of mesenchyme cells and topped by ectoderm, along which (antero-posterior, i.e. fore-to-aft) there is an 'apical ectodermal ridge' (AER). This arrangement of the bud continues to exist on the end of the limb as it elongates proximo-distally (i.e. outwards from the body). Work on developmental patterning of a limb commonly focuses on the skeletal structure, which forms first as cartilage and then becomes bone. The skeletal elements form sequentially from proximal to distal as the tip advances. To devotees of reaction–diffusion mechanisms it is obviously attractive to attach the sequence of one bone in the thigh (or upper arm) to two in the calf (or lower arm) to larger numbers of digits (according to species) to the changing fit of wavelength onto the discoid cross-section of the tip. The terminal bud has been given the name 'progress zone' on the basis that the mesenchyme cells in it advance in their developmental state while they are in the tip, in order to be at the right developmental level for the position at which they leave the progress zone and become 'frozen' in a particular state. This concept, for which grafting experiments (putting tips at one level onto limb stumps at a different level) provide evidence is in no way uniquely related to reaction–diffusion or any other mechanism for the patterning. It was, in fact, largely established in the first instance by people thinking in terms of the Wolpert 'positional information' concept, in the form that the time-sequence of cellular development in the progress zone leads to the sweeping out of a positional sequence along the limb (Wolpert 1971; Wolpert 1975; Tickle *et al.* 1975). At a late stage in this proximo-distal sequence, some kind of 'positional information' in the antero-posterior direction along the apical ectodermal ridge tells the system in what order to place the digits. This appears to be an exponential gradient of retinoic acid along the AER (Thaller and Eichele 1987). This and the Bcd protein distribution I have mentioned in *Drosophila* eggs are two quantitatively well-established exponential gradients of what I like to call 'Wolpert' or 'type I' morphogens, as distinct from 'Turing' or 'type II' morphogens that have to go in pairs. All-trans retinoic acid is a small (by biological standards) organic molecule closely related to vitamin A (retinol; which has $-CH_2OH$ instead of the terminal $-COOH$). The contrast between this and Bcd protein (55 kD) illustrates that Wolpert morphogens may be united only in the general nature of their developmental function, but very diverse in chemical structure. I expect the same to be established ultimately for Turing morphogens.

Reaction–diffusion has been advocated for vertebrate limb patterning, even as far as enrichment material for textbooks (item 16.1 on the website http://8e.devbio.com, a companion to Gilbert 2006). Speculative mechanisms, however, commonly include a selection of the things I have mentioned in connection with gastrulation and cell sorting (Sections 9.4 and 7.2.1): reaction–diffusion, mechanical forces and cell movement (individual or streaming), differential adhesion, and possible identities of some of the substances involved. Among these, animal tissues usually contain non-rigid proteinaceous structural components outside the cells, therefore known as extracellular matrix (ECM). These are predominantly collagens, to the extent that it has been said that if all the materials of a human body except collagens were to be removed, the body would still be visible, albeit in somewhat ghostly fashion, in its usual size and shape. (It has also been said that if all the materials of the world were removed except nematode worms, a ghostly world would still be seen as an astronomical body of the usual size and shape; perhaps fortunately, I cannot remember the authors to give an attribution for either of these statements.) Newman and Frisch (1979), Solursh *et al.* (1984) and Newman (1988) cite the ECM component fibronectin as a possibly important substance in limb patterning, perhaps even a Turing morphogen. Oster *et al.* (1983) give a thorough mechanochemical treatment, with features of cell movement towards cartilage-forming regions and reaction–diffusion-like instabilities that double up on the number of these regions, from one to two, where limb formation progresses from the upper to the lower limb. Newman (1996) gives a brief and readable review, wherein, nevertheless, he manages to list a lot of possible processes and quite an assortment of possibly significant substances including Hox gene products (DNA-binding proteins that can influence DNA activity and hence kinetic patterning) and gene products such as sonic hedgehog (Shh), WNTs and growth factors Fgf and Tgfβ, the same substances that are increasingly mentioned in connection with gastrulation, with WNT now having pride of place as first past the post to be identified definitely as a Turing activator for hair follicle formation (Sick *et al.* 2006).

9.5.2 Growing like plants?

I am looking for unities in the development of living things, and have therefore chosen to put plants and animals together in the same book. Where in this topic can we look across kingdom boundaries and see a familiar country on the other side? There are two things to look at: pattern-generating mechanisms, and the morphological changes they either initiate or continuously control. In biology in general, and developmental biology in particular, these are often carefully separated into 'pattern formation' and 'morphogenesis'. I find this particularly frustrating when it leads to all the talks I want to go to being lined up simultaneously in two parallel sessions. Nature took no such care in designing life. But the distinction between mechanisms and shape-changes can be useful in the enterprise of crossing boundaries. At the mechanistic level, I hope that I have demonstrated in the several accounts here that the same themes or motifs of reaction–diffusion can be sought (and sometimes even almost found!) in both animals and plants. A specific instance is that Holloway and I could read Meinhardt's (1995) sea-shell book and find in the 'extinguishing system' for a multiply repeated triangle pattern something that helped us in

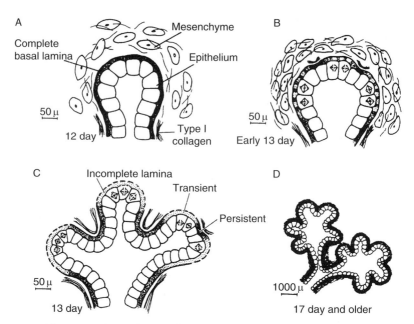

Figure 9.13 Stages in morphogenesis of the mouse submandibular salivary gland, to illustrate a type of morphogenesis in vertebrates somewhat resembling the common branching morphogenesis of plants. How is chemical activity linked to pattern changes in extracellular matrix (thinning at advancing lobe tips, thickening in notches), cell proliferation and changes in epithelial cell shapes? From Bernfield *et al.* (1984). © Wiley-Liss, with permission of Wiley-Liss, a division of John Wiley and Sons, Inc.

devising a mechanism for the doubly or triply repeated dichotomous branchings in *Micrasterias*.

At the morphological level, the execution of 'body plans' and the formation of many and diverse organs, often formed only once and in specific locations in the body, makes a big contrast between animal and plant development. For the latter, I have devoted a lot of space to branching processes. These can be found in particular kinds of animal organs having multiply lobed structures, e.g. lungs and various types of glands. Fig. 9.13 remains one of the best pictorial summaries of some candidates for the main protagonists in formation of such patterns, from the work of Bernfield *et al.* (1984) on the mouse submandibular salivary gland: Fig. 9.13A shows a lobe consisting of a sheet of epithelial cells covered by extracellular matrix in the form known as a 'basal lamina'. Outside this, there are mobile mesenchyme cells. Fig. 9.13B shows the lobe grows by cell divisions. Mitotic spindles are shown in some of the epithelial cells for divisions that would lead to area increase of the sheet in its own current plane. By itself, this could point towards computational modelling of shape change in which area increase is accommodated by outward movement, just like that described for plant growth in Chapter 5. But there is a lot more going on. Fig. 9.13C shows that where parts of the lobe are going to advance outwards, leading to branching, the basal lamina loses its integrity. But where the clefts between new lobes are going to be, it is persistent (compare to the differentially growing regions in Section 5.2.2 and Fig. 5.5). Also, not shown in B, C and D, the mesenchyme cells have a role in the morphogenesis. This seems to be more a matter of spatially patterned supply of mitogens manufactured by the mesenchyme cells (possibly reaction–diffusion) than of clustering of the cells themselves (which could involve mechanochemistry in the spirit of Oster *et al.* 1983). For lung branching morphogenesis, the website http://8e.devbio.com, accompanying Gilbert (2006), likewise mentions mesenchymal regulation by BMP4, Wnt2, Shh and Fgf10 proteins; compare substances mentioned for limb morphogenesis (Section 9.5.1), gastrulation (Section 9.4) and chick feather follicles (Section 6.3.5), for which Chuong strongly advocates Shh or Fgf4 as activator and BMP2 or BMP4 as inhibitor in a Turing morphogen pair.

Finally, a different aspect of growing like plants: segmentation, which I have discussed at some length for animals, but not at all for plants, except for a mention (Chapter 8) that the Japanese word *fushi* originally meant the repeated internodes on a bamboo stalk, an example of a common segmental feature of much plant development. The message for vertebrates is that they do not go about segmentation (somite formation in 'blocks', Section 9.3) in a manner anything like that of plants. But in passing, the annelid worms (earthworms, marine polychaete worms and leeches) have a mode of sequential production of segments that much resembles plant stalk development.

Epilogue

Lionel's text, to this point, brings us to 2008. Some further comments are therefore useful, both to round out Lionel's thoughts, and to relate his ideas more fully to current research in the field, now a rapidly expanding one. Lionel's unique voice is apparent throughout the book. It is not intended as a guide book to quantitative work in development – *Biological Physics of the Developing Embryo* by Forgacs and Newman (2005) is a good example of this. Nor is it a broad survey of pattern-formation processes and how they apply in biology – Philip Ball's *Nature's Patterns* (2009) is exemplary in that regard. Rather, the focus is on how one goes about exploring and testing the potential of a particular theory, developed in this case first by Turing and extended since by others; it is as much a commentary on how to construct a quantitative biology as it is an examination of particular dynamic issues in development. Lionel began his work in biology in the early 1970s, and was part of the blossoming of ideas in self-organization and how they might apply in biology. As chronicled here, these ideas gained a degree of acceptance by a segment of the developmental community, despite the division of cultures Lionel has described. The 1990s, though, heralded increasingly powerful techniques for manipulating gene regulation, with an increasingly detailed mapping of the components of developmental pathways, and little emphasis on overall dynamic constraints for patterning. These problems of spatial patterning had not gone away, however, and Lionel remained an evangelist for the cause. The idea for this book was originally as a more accessible version of his previous book, *Kinetic Theory of Living Pattern*, with a more general audience in mind. However, if the previous book was a summary of research and ideas through the 1980s, this book became more of a commentary on how to bridge the great divide between physical science and molecular biology as it was appearing in the late 1990s. And yet, during the writing of this book in the early 2000s, physical and mathematical approaches began to feature increasingly in mainstream biology, to the point that Lionel could feature early spans of the 'bridge', such as Sick *et al.* (2006).

What led to this new mathematical biology of the 2000s? To a degree, it relates to the arguments in this book, in that getting quantitative data (e.g. time-scales, concentrations, distances) is the key to testing any theory so that it can be usefully incorporated into mainstream research. But this has perhaps happened less because experimentalists became excited by theoretical issues, and more because the explosion of molecular biology data (see the Preface and Chapters 1 and 3) necessitated computational techniques (bioinformatics, systems biology). Experimentalists thus had to accommodate themselves to admitting mathematical methods into biology. While the inclusion of computation into mainstream biology was chiefly along the lines of '–omic' databases, of sequences and lists of molecular species, dynamic modelling also got pulled along for the ride. New models can take advantage of both the quantity of data and its high resolution. In many systems, theoreticians can now develop models with known molecules, and make predictions as to the concentrations of these molecules in time and space, and under various perturbations. The predictions can then be tested in the lab. It is a level of data resolution that has been hoped for since reaction–diffusion ideas were first being developed. At the same time, the greater acceptance of mathematical techniques in mainstream biology has helped foster the realization that many developmental phenomena, such as growth, are fundamentally quantitative, and that quantitative approaches are essential if they are to be fully understood (see Braybrook and Kuhlemeier (2010) for an excellent recent review).

There is perhaps a tendency as a new generation of science takes hold to emphasize a break between the new and the old. There is certainly a tone in the developmental modelling of the last ten years that a major step forward has been made from earlier theory that did not connect so well with molecular biologists. This book should demonstrate, though, that earlier theory did not aim to operate in a vacuum or be obtuse. Strong efforts have been made for decades to bridge quantitative theories and biology. Indeed, Lionel was one of the first to sit down and make the direct measurements needed to test patterning ideas. Now, thankfully, math is increasingly being applied to verbal molecular biology models. But the use of equations does not automatically create magic: useful modelling must always retain a focus on the questions being asked. Models must be formulated with particular variables to address particular problems, and they provide no new understanding if they do not provide answers in terms of what dynamic properties are responsible for particular phenomena. As data and computational power increase, there are efforts to include everything, to make an 'in silico cell' or an 'in silico embryo' (an analogue I have heard of in ecology is a 'model of California'). The dynamics of such models are likely to be too complex for responses to be interpretable: such a model might recapitulate an experimental result, but we would be unlikely to know why it did so. Newer models have the advantage over previous decades of having many more 'anchoring points' in experimental data, but models remain descriptive unless their behaviour is understood mathematically (i.e. in terms of the types of model solutions and their stability characteristics). With this, experiments can begin to distinguish between competing models. For instance: (1) To distinguish static versus dynamic (roughly Wolpert-type versus Turing-type) patterning mechanisms in *Drosophila* segmentation (Section 8.2), the behaviour of each needed to be characterized. Then, time-course data showed pattern shifts and pattern precision which could not be produced by a static mechanism; research is now focusing on the types of dynamic mechanisms involved. (2) In Chapter 2, the wavelength expressions for a number of different models are given. These not only indicate how models might be distinguished experimentally, but also demonstrate how a property such as wavelength is produced by a dynamic mechanism. (3) Mechanics and chemistry are both likely to operate in plant morphogenesis (Chapters 4 and 5). Models provide an understanding of the limits and hallmarks of each type of mechanism – experiments can investigate to what degree these are observed in particular phenomena, suggesting relative contributions from each mechanism. With each aspect so characterized, combined mechanochemical models can then be properly developed. There has been a recent push to develop 'multiscale' models, which combine multiple types of mechanism (chemical, mechanical, electrical, etc.) or different scales (molecular, cellular, organismal, etc.). These attempt to address very real biological complexities, but can contribute to theoretical understanding only insofar as the interactions between levels are understood mathematically. (See Rao and Arkin (2001) and Milo *et al.* (2002) for perspectives on combining dynamic 'modules' for developmental phenomena.) So, while the past decade has brought unprecedented coordination between models and experiments, the power of modelling is, as it always has been, in using mathematics to analyse potential mechanisms. At this fundamental level, there is no break between the 'old' and the 'new' developmental modelling – the core of the theory still lies in the mathematics. Major research streams continue on the dynamics themselves, allowing us to classify mechanisms, identify and distinguish them in biological systems, and begin to understand how they combine in complex phenomena. Such work produces the organizational principles of a developmental theory.

This book reminds us of the importance of focusing on dynamics and principles, and is replete with examples of how to do this. The breadth of organisms covered makes an additional point: much of the recent detailed matching between data and models is chiefly for selected 'model' species, such as *Drosophila* for insects and *Arabidopsis* for plants. In these organisms, the molecular tools exist to test model predictions for kinetics, transport and mechanics – to test for specific molecules in specific places and times. This has indeed advanced theory, validating some ideas and overturning others (see Jaeger (2009) for a comprehensive review in *Drosophila* segmentation, cf. Chapter 8), and these can certainly have implications beyond the model species. But quantitative experiments are difficult, and still the exception to the rule even in model organisms (though see http://urchin.spbcas.ru/flyex and http://bdtnp.lbl.gov/Fly-Net for spatially and temporally quantitative atlases of *Drosophila* segmentation patterns). More to the point, there are limits to the phenomena that can be studied with model organisms. Outside these limits, indirect non-molecular approaches may provide the best options. For instance, the control of conifer cotyledon spacing explored in Chapter 3 is not something that can be studied in the fixed-number monocotyledonous and dicotyledonous model plants – yet conifers and their development are not unimportant (perhaps especially in Canada!). Understanding growth is another case – molecular tools for affecting cell mechanics and cell expansion are advancing, but much macroscopic work is still needed to understand expansion rates and what these imply for physical and chemical models (e.g. Dumais *et al.* 2004). Even in model species, molecular biology will not provide the complete answer to problems such as these (and model species chosen for molecular manipulation aren't necessarily the best suited for such macroscopic investigations). Model species can provide critical advances in particular areas, but don't cover a whole range of developmental questions. Quantitative analysis without reference to specific molecules can be immensely powerful in this broader context; Lionel's thermo-dynamic approach with *Acetabularia* (Chapter 3) remains a singular illustration of this fact.

In what areas is theory advancing the understanding of developmental principles now? Perhaps three broad areas can extend what has been discussed in this book: chemical pattern formation; shape change; and noise. Pattern formation, how chemicals get in the right concentrations in the right places, is at the heart of this book. Reaction–diffusion continues to be a strong field, as indicated in Fig. 6.4. A special issue of the *International Journal of Developmental Biology* (2009, vol. 53, no. 5/6; online) gives a recent overview. Skin patterning (Sick *et al.* 2006) and plant leaf trichome patterning (Digiuni *et al.* 2008) are excellent cases of direct molecular-biological determination of Turing patterns in vivo. Reaction–diffusion has also been tied very closely to molecular data in single cells: a number of authors have recently used reaction–diffusion to address pattern formation in *E. coli* bacteria, where the MinD/MinE proteins concentrate at different halves of the cell and determine the location of cell division (Meinhardt and de Boer 2001; Howard *et al.* 2001; Hunding 2004; Howard and Kruse 2005). Work in the past decade has shown a number of dynamic motifs that operate in spatial and temporal patterning; Turing patterning can be considered as a class within broader interactions which include feedback, feedforward, etc., and can give rise to very rich and non-linear temporal or spatial behaviour. These motifs are becoming commonly accepted as determining many cases of developmental patterning. Examples include the repressilator motif (Elowitz and Leibler 2000) and bistability (e.g. Guidi and Goldbeter 2000; Wang and Ferguson 2005; Umulis *et al.* 2006; Lopes *et al.* 2008). For recent summaries and discussions, see Papatsenko (2009) and Isalan (2009); especially for the limitations of binary logic networks, in comparison with kinetic models, in modelling developmental phenomena). Transport, key to reaction–diffusion patterning, is also broadening, with detailed work on the formation of monotonic gradients providing new insight into the role of active transport mechanisms (as opposed to passive diffusion). A number of authors have made very quantitative advances in extracellular gradient formation and signalling in dorso-ventral patterning and patterning within the wing in *Drosophila* (Lander *et al.* 2002; Eldar *et al.* 2003; Mizutani *et al.* 2005; Bollenbach *et al.* 2005, 2008; Umulis *et al.* 2010; reviews in Lander 2007; Reeves *et al.* 2007; and Lander *et al.* 2009). In antero-posterior patterning, as touched on in Chapter 8, new experiments (e.g. Gregor *et al.* 2007a, 2007b; Weil *et al.* 2006, 2008; Spirov *et al.* 2009) and models (e.g. Coppey *et al.* 2007, 2008) are finding fundamental roles for active transport along cytoskeletal tracks, and also finding trapping, by nuclei or cytoskeleton, to be important in gradient formation. It is very promising that mathematical analysis is being applied to a number of models that have been built up from detailed biological data, in order to understand the underlying dynamics (e.g. Muratov and Shvartsman 2003; Gursky *et al.* 2006; Manu *et al.* 2009b). Further overviews are offered in two recent (2010) books: *Symmetry Breaking in Biology* (Li and Bowerman, 2010) and *Generation and Interpretation of Morphogen Gradients* (Briscoe *et al.* 2010).

How organisms generate shape goes to the title of this book, and Lionel, in keeping with Turing, rarely wanted to think of chemical patterning in isolation from shape change: the impetus for much of his work was to explore the symmetry-breaking afforded by chemical patterning as the precursor and driver of shape change in tissues. As discussed in this book, this coupling of patterning, growth and shape change is most evident in plant development, though it is critical in many events in animal development. Chapter 5 goes directly to our work coupling reaction–diffusion dynamics to morphogenesis. A number of other theoreticians are now also addressing the dynamics of pattern-growth coupling, such as Crampin *et al.* (2002), Salazar-Ciudad *et al.* (2003), Salazar-Ciudad and Jernvall (2004), and Neville *et al.* (2006). Within plant modelling, there have been impressive 'new' mathematical biology breakthroughs in the past decade, chiefly for the model plant *Arabidopsis* (with a major proliferation of departments and centres devoted to plant systems biology worldwide). Earlier ideas of Mitchison (1980, 1981) on transport of the hormone auxin have been verified and expanded through both experiments that have uncovered the molecular details of how auxin is transported across cells (e.g. Reinhardt *et al.* 2003; Benkova *et al.* 2003) and models of these processes, which can generate phyllotactic patterns for leaf initiation (e.g. Smith *et al.* 2006; de Reuille *et al.* 2006; Jönsson *et al.* 2006). Auxin patterning has also been coupled to morphogenesis at high resolution in *Arabidopsis* roots (Grieneisen *et al.* 2007; Laskowski *et al.* 2008). As introduced in Chapter 4, there is a very large community investigating the mechanical properties of cells and tissues, especially with respect to plant morphogenesis. A great deal of work lies ahead both in elucidating the mechanics and in coupling these to chemical patterning mechanisms: plant development uses both. Work advances at both the cellular level (e.g. Hamant *et al.* 2008) and at the tissue continuum level (e.g. Rolland-Lagan *et al.* 2003; Dumais *et al.* 2004). Within the context of quantitative molecular biology models and mechanical models, reaction–diffusion has much to offer as a route to the dynamics: minimal models for pattern formation (such as the Brusselator), representing dynamics found in larger biochemical networks, can readily be coupled to other processes, such as growth and shape change, and understood mathematically. Molecular data can easily be incorporated into such model frameworks, as well as additional mechanisms (e.g. mechanics) for growth and shape change. A key component of morphogenesis is pattern selection – either from an unpatterned state, or from pattern to pattern in concert with growth (as in Chapter 5). Reaction–diffusion theory is well-developed to make key contributions in this area of morphogenesis. These issues are explored further in Holloway (2010).

The control of natural fluctuations is an issue discussed to some degree in Chapter 8, and which Lionel was involved in since the early 1990s (see Lacalli and Harrison 1991). How much do developmental outcomes vary in a natural population? How is the low variability seen in most developmental patterns achieved by mechanisms that operate at low concentrations, for which basic chemistry would predict very high noise? Developmental reliability is a fundamental biological question, and a quantitative approach is required: mathematics and statistics are necessary for both analysing data and developing theory. Noise can arise, and be controlled, at multiple levels, from the random nature of reaction and transport events, to the random aspects of an organism's inheritance, to its linkage with its environment. In the past decade, studies in single-celled organisms (*E. coli* and yeast), combining quantitative measurements of gene expression with mathematical models, have greatly deepened the understanding of how noise arises, is propagated, and can be reduced (for example, see Rao *et al.* 2002; Blake *et al.* 2003; Pedraza and van Oudenaarden 2005; Rosenfeld *et al.* 2005; Isaacs *et al.* 2005; Kaern *et al.* 2005; Colman-Lerner *et al.* 2005). Noise can give rise to diversity in phenotypes, which is crucial to evolution (e.g. Ko 1991, 1992; Elowitz *et al.* 2002); but within embryos, noise carries the large potential to disrupt developmental programmes. Broad-scale comparison of relative noise between genes can give insight into where in the molecular machinery noise reduction may be most critical (e.g. Fraser *et al.* 2004; Bar-Even *et al.* 2006). More specific to the topic of this book, the understanding of what makes spatial patterns reliable is still at an early stage. In *Drosophila* (Chapter 8), a number of studies have begun to characterize variability between embryos (e.g. Houchmandzadeh *et al.* 2002; Holloway *et al.* 2006; Gregor *et al.* 2007a, 2007b; Lott *et al.* 2007; He *et al.* 2008). Studying variability within tissues presents more challenges for separating experimental from intrinsic noise. Some data have been presented in Spirov and Holloway (2003), Gregor *et al.* (2007a, 2007b), Wu *et al.* (2007) and Alexandrov *et al.* (2008). Theory for the generation and reduction of such internal noise has been explored by Spirov *et al.* (2008), Tkacik *et al.* (2008) and Okabe-Oho *et al.*

(2009). Many of the dynamic mechanisms for noise reduction discovered in the unicellular work may apply to reliability of spatial patterns. Within the theme of this book, though, Turing reaction–diffusion has a direct capacity for reducing variability in spatial pattern: because of the 'band-pass' selection of pattern elements (Section 6.3.3), Turing patterns are very robust to noise. This was introduced in Section 8.3 (Gierer and Meinhardt 1972; Holloway and Harrison 1999b), but the concept that pattern is a distinct entity that has de-selected other potential patterns, including noise (e.g. short wavelengths), should lie at the heart of future explorations of how embryos ensure the reliable formation of their spatial patterns.

And what of Lionel's final research projects? There are two: the first is a continuation of the chromatin patterning project discussed in Section 7.2.2. Newer experimental techniques are upholding and enriching the proposed spinodal decomposition mechanism. These advances have recently been published in Martens *et al.* (2009). And, Richard Adams worked with Lionel for most of the past decade on developing a purely chemical mechanism that can control the boundaries of pattern formation in a growing system – the double Brusselator discussed in Section 5.4. An article on this work is in preparation.

Mathematical techniques are being incorporated into developmental biology as never before; the science will become quantitative. What is perhaps emerging as a key point of developing and testing models at the molecular level (the 'new' mathematical biology), is that it is making less and less sense to talk about the application of physics and chemistry to biology: rather we are increasingly appreciating the unique dynamics inherent in biological systems – levels of complexity and modes of organization that were never conceived in non-biological systems. And, aided by comparative biology at the DNA-sequence level, we are beginning to appreciate how these developmental modes have arisen through evolutionary processes. This will be the true coming of age for the field, when we can begin to appreciate the scope and depth of life from a single discipline, rather than across a gulf between disciplines.

References

Aida, M., Ishida, T., Fukaki, H., Fujisawa, H. and Tasaka, M. (1997). Genes involved in organ separation in *Arabidopsis*: an analysis of *cup-shaped cotyledon* mutant. *Plant Cell* **9**, 841–57.

Akam, M. E. (1987). The molecular basis for metameric pattern in the *Drosophila* embryo. *Development* **101**, 1–22.

Akam, M. E. (1989). Making stripes inelegantly. *Nature* **341**, 282–3.

Alexandrov, T., Golyandina, N. and Spirov, A. V. (2008). Singular spectrum analysis of gene expression profiles of the early *Drosophila* embryo: exponential-in-distance patterns. *Res. Lett. in Signal Processing*, **12**, n.p.

Amtmann, A., Klieber, H. G. and Gradmann, D. (1992). Cytoplasmic free Ca^{2+} in the marine alga *Acetabularia*: measurement with Ca^{2+}-selective microelectrodes and kinetic analysis. *J. Exp. Bot.* **43**, 875–85.

Angier, N. (2007). *The Canon*. Boston and New York: Houghton Mifflin.

Armstrong, J. B. (1989). A Turing model to explain heart development. *Axolotl Newsletter* **18**, 23–5.

Armstrong, J. B. and Graveson, A. C. (1988). Progressive patterning precedes segmentation in the Mexican axolotl (*Ambystoma mexicanum*). *Dev. Biol.* **126**, 1–6.

Armstrong, J. B. and Malacinski, G. M. (1989). *Developmental Biology of the Axolotl*. New York: Oxford University Press.

Baker, R. E., Schnell, S. and Maini, P. K. (2009). Waves and patterning in developmental biology: vertebrate segmentation and feather bud formation as case studies. *Int. J. Dev. Biol.* **53**, 783–94.

Ball, P. (2009). *Nature's Patterns: A Tapestry in Three Parts*. Oxford: Oxford University Press.

Bar-Even, A., Paulsson, J., Maheshri, N., Carmi, M., O'Shea, E., Pilpel, Y. and Barkai, N. (2006). Noise in protein expression scales with natural protein abundance. *Nat. Genetics* **38**, 636–43.

Barlow, P. W. (1984). Positional controls in root development. In *Positional Controls in Plant Development*, ed. P. W. Barlow and D. J. Carr, pp. 281–318. Cambridge: Cambridge University Press.

Battey, N. H. and Blackbourn, H. D. (1993). The control of exocytosis in plant cells. *New Phytol.* **125**, 307–38.

Belousov, B. P. (1959). A periodic reaction and its mechanism. In *Collection of Short Papers on Radiation Medicine for 1958*. Moscow: Med. Publ. Reprinted In *Oscillations and Traveling Waves in Chemical Systems*, eds R. J. Field and M. Burger. New York: Wiley, 1985.

Benková, E., Michniewicz, M., Sauer, M., Teichman, T., Seifertova, D., Jurgens, G. and Friml, J. (2003). Local, efflux-dependent auxin gradients as a common module for plant organ formation. *Cell* **115**, 591–602.

Berger, S. and Kaever, M.J. (1992). *Dasycladales: An Illustrated Monograph of a Fascinating Algal Order*. Stuttgart and New York: Thieme.

Bergmann, S., Sandler, O., Sberro, H., Schejter, E., Shilo, B.-Z. and Barkai, N. (2007). Pre steady-state decoding of the Bicoid morphogen gradient. *PLoS Biology* **5**, e46.

Bernfield, M., Banerjee, S.D., Koda, J.E. and Rapraeger, A.C. (1984). Remodelling of the basement membrane as a mechanism of morphogenetic tissue inter-action. In *The Role of Extracellular Matrix in Development*, ed. R.L. Trelstad, pp. 545–71. New York: Alan R. Liss.

Blake, W.J., Kærn, M., Cantor, C.R. and Collins, J.J. (2003). Noise in eukaryotic gene expression. *Nature* **422**, 633–7.

Bohn, S., Andreotti, B., Douady, S., Munzinger, J. and Couder, Y. (2002). Constitutive property of the local organization of leaf venation networks. *Phys. Rev.* **E65**, 061914.

Bollenbach, T., Kruse, K., Pantazis, P., González-Gaitán, M. and Jüllicher, F. (2005). Robust formation of morphogen gradients. *Phys. Rev. Lett.* **94**, 018103.

Bollenbach, T., Pantazis, P., Kicheva, A., Bokel, C., González-Gaitán, M. and Jüllicher, F. (2008). Precision of the Dpp gradient. *Development* **135**, 1137–46.

Bonotto, S., and Berger, S. (1997). *Proceeding/Symposium Ecology and Biology of Giant Unicellular Algae*. Torino: Museo regionale di scienze naturali.

Borckmans, P., Dewel, G., De Wit, A. and Walgraef, D. (1995). Turing bifurcations and pattern selection. In *Chemical Waves and Patterns*, eds. R. Kapral and K. Showalter, pp. 323–63. Dordrecht: Kluwer.

Bornholdt, S. (2005). Less is more in modeling large genetic networks. *Science* **310**, 449–51.

Braybrook, S. and Kuhlemeier, C. (2010). How a plant builds leaves. *Plant Cell* **22**, (in press).

Brière, C. and Goodwin, B.C. (1988). Geometry and dynamics of tip morphogen-esis in *Acetabularia. J. Theor. Biol.* **131**, 461–75.

Briscoe, J., Lawrence, P.A., and Vincent, J-P., eds. (2010). *Generation and Interpret-ation of Morphogen Gradients*. Cold Spring Harbor: Cold Spring Harbor Laboratory Press.

Byrne, G. and Cox, E.C. (1986). Spatial patterning in *Polysphondylium*: monoclonal antibodies specific for whorl prepatterns. *Dev. Biol.* **117**, 442–55.

Byrne, G. and Cox, E.C. (1987). Genesis of a spatial pattern in the cellular slime mold *Polysphondylium pallidum. Proc. Nat. Acad. Sci. USA* **84**, 4140–4.

Cahn, J.W. (1965). Phase separation by spinodal decomposition in isotropic systems. *J. Chem. Phys.* **42**, 93–9.

Cahn, J.W. and Hilliard, J.E. (1958). Free energy of a nonuniform system: I. Interfacial energy. *J. Chem. Phys* **28**, 258–67.

Castets, V., Dulos, E., Boissonade, J. and de Kepper, P. (1990). Experimental evidence of a sustained standing Turing-type nonequilibrium chemical pat-tern. *Phys. Rev. Lett.* **64**, 2953–6.

Chadefaud, M. (1952). La leçon des algues: comment elles ont évolué; comment leur évolution peut éclairer celle des plantes supérieures. *L'Année Biologique* **28**, C9–C25.

Chevaillier, P. (1970). Le noyau du spermatozöide et son évolution au cours de la spermiogenèse. In *Comparative Spermatology*, Vol. **5**, ed. B. Baccetti, pp. 499–514. New York: Academic Press.

Church, A.H. (1895). The structure of the thallus of *Neomeris dumetosa*, Lamour. *Ann. Bot.* **9**, 581–608.

Church, A.H. (1919). Thalassiophyta and the subaerial transmigration. *Botanical Memoirs* **3**, 1–95.

Claxton, J.H. (1964). The determination of patterns with special reference to that of the central primary skin follicles in sheep. *J. Theor. Biol.* **7**, 302–17.

Clyde, D., Corado, M., Wu, X., Pare, A., Papatsenko, D. and Small, S. (2003). A self-organizing system of repressor gradients establishes segmental complexity in *Drosophila*. *Nature* **426**, 849–53.

Colman-Lerner, A., Gordon, A., Serra, E., Chin, T., Resnekov, O., Endy, D., Pesce, C.G. and Brent, R. (2005). Regulated cell-to-cell variation in a cell-fate decision system. *Nature* **437**, 699–706.

Coppey, M., Berezhovskii, A.M., Kim, Y., Boettiger, A.N. and Shvartsman, S.Y. (2007). Modeling the bicoid gradient: diffusion and reversible nuclear trapping of a stable protein. *Dev. Biol.* **312**, 623–30.

Coppey, M., Boettiger, A.N., Berezhovskii, A.M. and Shvartsman, S.Y. (2008). Nuclear trapping shapes the terminal gradient in the *Drosophila* embryo. *Curr. Biol.* **18**, 915–19.

Couté, A. and Tell, G. (1981). *Ultrastructure de la paroi cellulaire des Desmidiacees au microscope electronique a balayage*. Vaduz: J. Cramer.

Crampin, E.J., Hackborn, W.W. and Maini, P.K. (2002). Pattern formation in reaction–diffusion models with nonuniform domain growth. *Bull. Math. Biol.* **64**, 747–69.

Crauk, O. and Dostatni, N. (2005). Bicoid determines sharp and precise target gene expression in the *Drosophila* embryo. *Curr. Biol.* **15**, 1888–98.

Crombie, A.C. (1959). *Medieval and Early Modern Science*, 2nd ed., **2** vols. Garden City, N Y: Doubleday.

Damen, W.G.M. (2007). Evolutionary conservation and divergence of the segmentation process in arthropods. *Dev. Dyn.* **236**, 1379–91.

Davis, G.K. and Patel N.H. (1999). The origin and evolution of segmentation. *Trends Genet.* **15**, M68–M72.

Davis, G.K., Jaramillo, C.A. and Patel, N.H. (2001). Pax group III genes and the evolution of insect pair-rule patterning. *Development* **128**, 3445–58.

Degn, H. (1972). Oscillating chemical reactions in homogeneous phase. *J. Chem. Ed.* **49**, 302–7.

de Reuille P.B., Bohn-Courseau, I., Ljung, K., *et al.* (2006). Computer simulations reveal properties of the cell–cell signalling network at the shoot apex in *Arabidopsis*. *Proc. Nat. Acad. Sci. USA* **103**, 1627–32.

De Robertis, E.M. (2008). The molecular ancestry of segmentation mechanisms. *Proc. Nat. Acad. Sci. USA* **105**, 16411–12.

Digiuni, S., Schellmann, S., Geier, F., *et al.* (2008). A competitive complex formation mechanism underlies trichome patterning on *Arabidopsis* leaves. *Mol. Sys. Biol.* **4**, 217.

Douady, S. and Couder, Y. (1996). Phyllotaxis as a dynamical self organizing process. I, II, III. *J. Theor. Biol.* **178**, 255–312.

Driever, W. and Nüsslein-Volhard, C. (1988). A gradient of *bicoid* protein in *Drosophila* embryos: the *bicoid* protein determines position in the *Drosophila* embryo in a concentration-dependent manner. *Cell* **54**, 83–93, 95–104.

Dumais, J. and Harrison, L.G. (2000). Whorl morphogenesis in the dasycladalean algae: the pattern formation viewpoint. *Phil. Trans. R. Soc. Lond. B* **355**, 281–305.

Dumais, J. and Steele, C. (2000). New evidence for the role of mechanical forces in the shoot apical meristem. *J. Plant Growth Regulation* **19**, 7–18.

Dumais, J., Long, S.R. and Shaw, S.L. (2004). The mechanics of surface expansion anisotropy in *Medicago truncatula* root hairs. *Plant Physiol.* **136**, 3266–75.

Dumais, J., Shaw, S. L., Steele, C. R., Long, S. R. and Ray, P. M. (2006). An anisotropic-viscoplastic model of plant cell morphogenesis by tip growth. *Int. J. Dev. Biol.* **50**, 209–22.

Easton, H. S., Armstrong, J. B. and Smith, S. C. (1994). Heart specification in the Mexican axolotl (*Ambystoma mexicanum*). *Dev. Dyn.* **200**, 313–20.

Edelstein-Keshet, L. (1988). *Mathematical Models in Biology*. New York: Random House.

Eldar, A., Rosin, D., Shilo, B.-Z. and Barkai, N. (2003). Self-enhanced ligand degradation underlies robustness of morphogen gradients. *Dev. Cell* **5**, 635–46.

Elowitz, M. B. and Leibler, S. (2000). A synthetic oscillatory network of transcriptional regulators. *Nature* **403**, 335–8.

Elowitz, M. B., Levine, A. J., Siggia, E. D. and Swain, P. S. (2002). Stochastic gene expression in a single cell. *Science* **297**, 1183–6.

Emberger, L. (1968). *Les plantes fossiles dans leurs rapports avec les végétaux vivants*. Paris: Masson.

Ermentrout, B. (1991). Stripes or spots? Nonlinear effects in bifurcation of reaction–diffusion equations on the square. *Proc. R. Soc. Lond. A* **434**, 413–17.

Field, R. J., Körös, E. and Noyes, R. M. (1972). Oscillations in chemical systems: II. Thorough analysis of temporal oscillations in the bromate–cerium–malonate system. *J. Am. Chem. Soc.* **94**, 8649–64.

Forgacs, G. and Newman, S. A. (2005). *Biological Physics of the Developing Embryo*. Cambridge: Cambridge University Press.

Foty, R. A. and Steinberg, M. S. (2005). The differential adhesion hypothesis: a direct evaluation. *Dev. Biol.* **278**, 255–63.

Foty, R. A., Pfleger, C. M., Forgacs, G. and Steinberg, M. S. (1996). Surface tensions of embryonic cells predict their mutual envelopment behavior. *Development* **122**, 1611–20.

Fowlkes, C. C., Luengo Hendriks, C. L., Keränen, S. V. E., *et al.* (2008). A quantitative spatio-temporal atlas of gene expression in the *Drosophila* blastoderm. *Cell* **133**, 364–74.

Frank, F. C. (1953). On spontaneous asymmetric synthesis. *Biochim. Biophys. Acta* **11**, 459–63.

Frankel, J. (1989). *Pattern Formation: Ciliate Studies and Models*. Oxford: Oxford University Press.

Frasch, M., Hoey, T., Rushlow, C., Doyle, H. and Levine, M. (1987). Characterization and localization of the *even-skipped* protein of *Drosophila*. *EMBO J.* **6**, 749–59.

Fraser, H. B., Hirsh, A. E., Giaever, G., Kumm, J. and Eisen, M. B. (2004). Noise minimization in eukaryotic gene expression. *PLoS Biology* **2**, 834–838.

French, V., Bryant, P. J. and Bryant, S. V. (1976). Pattern regulation in epimorphic fields. *Science* **193**, 969–81.

Gardner, M. (1982). *The Ambidextrous Universe*, 2nd ed. Harmondsworth: Penguin Books.

Gibor, A. (1966). *Acetabularia*: a useful giant cell. *Sci. Am.* **215**, 118–24.

Giddings, T. H., Brower, D. L. and Staehelin, L. A. (1980). Visualization of particle complexes in the plasma membrane of *Micrasterias denticulata* associated with the formation of cellulose fibrils in primary and secondary cell walls. *J. Cell Biol.* **84**, 327–39.

Gierer, A. and Meinhardt, H. (1972). A theory of biological pattern formation. *Kybernetik* **12**, 30–9.

Gilbert, S. F. (2006). *Developmental Biology*, 8th ed. Sunderland, MA: Sinauer Associates.

Gilbert, S. F. and Sarkar, S. (2000). Embracing complexity: organicism for the twenty-first century. *Dev. Dyn.* **219**, 1–9.

Glansdorff, P. and Prigogine, I. (1971). *Thermodynamic Theory of Structure, Stability, and Fluctuations*. New York: Wiley-Interscience.

Glazier, J.A. and Graner, F. (1993). Simulation of the differential adhesion driven rearrangement of biological cells. *Phys. Rev. E* **47**, 2128–54.

Goel, N.S. and Rogers, G. (1978). Computer simulations of engulfment and other movements of embryonic tissues. *J. Theor. Biol.* **71**, 103–40.

Goodwin, B.C. and Trainor, L.E.H. (1985). Tip and whorl morphogenesis in *Acetabularia* by calcium-regulated strain fields. *J. Theor. Biol.* **117**, 79–106.

Goriely, A. and Tabor, M. (2003). Biomechanical models of hyphal growth in actinomycetes. *J. Theor. Biol.* **222**, 211–18.

Graner, F. and Glazier, J.A. (1992). Simulation of biological cell sorting using a two-dimensional extended Potts model. *Phys. Rev. Lett.* **69**, 2013–16.

Gratzl, M. (1980). Transport of membranes and vesicle contents during exocytosis. In *Biological Chemistry of Organelle Formation*, eds T. Bücher, W. Seebald, and H. Weiss, pp. 165–74. Berlin, Heidelberg and New York: Springer.

Green, P.B. (1999). Expression of pattern in plants: combining molecular and calculus-based biophysical paradigms. *Am. J. Bot.* **86**, 1059–76.

Green, P.B. and King, A. (1966). A mechanism for the origin of specifically oriented textures in development with special reference to *Nitella* wall texture. *Aust. J. Biol. Sci.* **19**, 421–37.

Green, P.B. and Linstead, P. (1990). A procedure for SEM of complex shoot structures applied to the inflorescence of snapdragon (*Antirrhinum*). *Protoplasma* **158**, 33–8.

Green, P.B., Steele, C.R. and Rennich, S.C. (1996). Phyllotactic patterns: a biophysical mechanism for their origin. *Ann. Bot.* **77**, 515–27.

Gregor, T., Bialek, W., van Steveninck, R.R.R., Tank, D.W. and Wieschaus, E.F. (2005). Diffusion and scaling during early embryonic pattern formation. *Proc. Nat. Acad. Sci. USA* **102**, 18403–7.

Gregor, T., McGregor, A.P. and Wieschaus, E.W. (2008). Shape and function of the bicoid morphogen gradient in dipteran species with different sized embryos. *Dev. Biol.* **316**, 350–8.

Gregor, T., Tank, D.W., Wieschaus, E.F. and Bialek, W. (2007a). Probing the limits to positional information. *Cell* **130**, 153–64.

Gregor, T., Wieschaus, E.F., McGregor, A.P., Bialek, W. and Tank, D.W. (2007b). Stability and nuclear dynamics of the bicoid morphogen gradient. *Cell* **130**, 141–52.

Grieneisen, V.A., Xu, J., Maree, A.F.M., Hogeweg, P. and Scheres, B. (2007). Auxin transport is sufficient to generate a maximum and gradient guiding root growth. *Nature* **449**, 1008–13.

Gross, J.D., Peacey, M.J. and von Strandmann, R.P. (1988). Plasma membrane proton pump inhibition and stalk cell differentiation in *Dictyostelium discoideum*. *Differentiation* **38**, 91–8.

Guidi, G.M. and Goldbeter, A. (2000). Oscillations and bistability predicted by a model for a cyclical bienzymatic system involving the regulated isocitrate dehydrogenase reaction. *Biophys. Chem.* **83**, 153–70.

Gunning, B.E.S. (1981). Microtubules and cytomorphogenesis in a developing organ: the root primordium of *Azolla pinnata*. In *Cytomorphogenesis in Plants*, ed. O. Kiermayer, pp. 301–26. Vienna: Springer-Verlag.

Gunning, B.E.S. (1982). The root of the water fern *Azolla*: cellular basis of development and multiple roles for cortical microtubules. In *Developmental Order: Its Origin and Regulation*, ed. S. Subtelny and P.B. Green, pp. 379–421. New York: Alan R. Liss.

Gursky, V.V., Kozlov, K.N., Samsonov, A.M. and Reinitz, J. (2006). Cell divisions as a mechanism for selection in stable steady states of multi-stationary gene circuits. *Physica D* **218**, 70–6.

Hafen, E., Kuroiwa, A. and Gehring, W.J. (1984). Spatial distribution of transcripts from the segmentation gene *fushi tarazu* during *Drosophila* embryonic development. *Cell* **37**, 833–41.

Hagemann, W. (1992). The relationship of anatomy to morphology in plants: a new theoretical perspective. *Int. J. Plant Sci.* **153**, S38–S48.

Hamant, O., Heisler, M.G., Jönsson, H., *et al.* (2008). Developmental patterning by mechanical signals in *Arabidopsis*. *Science* **322**, 1650–5.

Hardway, H., Mukjopadhyay, B., Burke, T., Hichman, T.J. and Forman, R. (2008). Modeling the precision and robustness of the Hunchback boundary during *Drosophila* embryonic development. *J. Theor. Biol.* **254**, 390–9.

Harrison, L.G. (1973). Evolution of biochemical systems with specific chirality: a model involving territorial behaviour. *J. Theor. Biol.* **39**, 333–41.

Harrison, L.G. (1974). The possibility of spontaneous resolution of enantiomers on a catalyst surface. *J. Mol. Evol.* **4**, 99–111.

Harrison, L.G. (1979). Molecular asymmetry and morphology: big hands from little hands. In *Origins of Optical Activity in Nature*, ed. D.C. Walker, pp. 125–40. Amsterdam: Elsevier.

Harrison, L.G. (1981). Physical chemistry of biological morphogenesis. *Chem. Soc. Rev.* **10**, 491–528.

Harrison, L.G. (1982). An overview of kinetic theory in developmental modelling. In *Developmental Order: Its Origin and Regulation*, ed. S. Subtelny and P.B. Green, pp. 3–33. New York: Alan R. Liss.

Harrison, L.G. (1992). Reaction–diffusion theory and intracellular differentiation. *Int. J. Plant Sci.* **153**, S76–S85.

Harrison, L.G. (1993). *Kinetic Theory of Living Pattern*. Cambridge: Cambridge University Press.

Harrison, L.G. (1994). Kinetic theory of living pattern. *Endeavour* **18**, 130–6.

Harrison, L.G. and Hillier, N.A. (1985). Quantitative control of *Acetabularia* morphogenesis by extracellular calcium: a test of kinetic theory. *J. Theor. Biol.* **114**, 177–92.

Harrison, L.G. and Kolář, M. (1988). Coupling between reaction–diffusion prepattern and expressed morphogenesis. *J. Theor. Biol.* **130**, 493–515.

Harrison, L.G. and Lacalli, T.C. (1978). Hyperchirality: a mathematically convenient and biochemically possible model for the kinetics of morphogenesis. *Proc. R. Soc. Lond.* **B202**, 361–97.

Harrison, L.G. and Tan, K.Y. (1988). Where may reaction–diffusion mechanisms be operating in metameric patterning of *Drosophila* embryos? *BioEssays* **8**, 118–24.

Harrison, L.G. and von Aderkas, P. (2004). Spatially quantitative control of the number of cotyledons in a clonal population of somatic embryo of hybrid larch *Larix* X *leptoeuropaea*. *Ann. Bot.* **93**, 423–34.

Harrison, L.G., Graham, K.T. and Lakowski, B.C. (1988). Calcium localization during *Acetabularia* whorl formation: evidence supporting a two-stage hierarchical mechanism. *Development* **104**, 255–62.

Harrison, L.G., Holloway, D.M. and Orchard, J.J. (1996). Hearts and somites: problems of pattern formation in vertebrate embryology modelled by reaction–diffusion. *Dev. Biol.* **175**, 73.

Harrison, L.G., Wehner, S. and Holloway, D.M. (2001). Complex morphogenesis of surfaces: theory and experiment on coupling of reaction–diffusion to growth. *Faraday Disc.* **120**, 277–94.

Harrison, L. G., Kasinsky, H. E., Ribes, E. and Chiva, M. (2005). Possible mechanisms for early and intermediate stages of sperm chromatin condensation patterning involving phase separation dynamics. *J. Exp. Zool.* **303A**, 76–92.

Harrison, L. G., Snell, J., Verdi, R., Vogt, D. E., Zeiss, G. D. and Green, B. R. (1981). Hair morphogenesis in *Acetabularia mediterranea*: temperature-dependent spacing and models of morphogen waves. *Protoplasma* **106**, 211–21.

Harrison, L. G., Donaldson, G., Lau, W., *et al.* (1997). CaEGTA uncompetitively inhibits calcium activation of whorl morphogenesis in *Acetabularia*. *Protoplasma* **196**, 190–6.

Hartmann, C., Taubert, H., Jäckle, H. and Pankratz, M. J. (1994). A 2-step mode of stripe formation in the *Drosophila* blastoderm requires interactions among primary pair-rule genes. *Mech. of Dev.* **45**, 3–13.

He, F., Wen, Y., Lin, X., *et al.* (2008). Probing intrinsic properties of a robust morphogen gradient in *Drosophila*. *Dev. Cell* **15**, 558–67.

Herschkowitz-Kaufman, M. (1975). Bifurcation analysis of nonlinear reaction-diffusion equations: II. Steady-state solutions and comparison with numerical simulations. *Bull. Math. Biol.* **37**, 589–636.

Holloway, D. M. (1995). *Reaction–Diffusion Theory of Localized Structures with Application to Vertebrate Organogenesis.* PhD thesis, University of British Columbia, Canada.

Holloway, D. M. (2010). The role of chemical dynamics in plant morphogenesis. *Bioch. Soc. Trans.* **38**, 645–50.

Holloway, D. M. and Harrison, L. G. (1995). Order and localization in reaction-diffusion pattern. *Physica A* **222**, 210–33.

Holloway, D. M. and Harrison, L. G. (1999a). Algal morphogenesis: modelling interspecific variation in *Micrasterias* with reaction–diffusion patterned catalysis of cell surface growth. *Phil. Trans. R. Soc. Lond. B* **354**, 417–33.

Holloway, D. M. and Harrison, L. G. (1999b). Suppression of positional errors in biological development. *Math. Biosc.* **156**, 271–90.

Holloway, D. M. and Harrison, L. G. (2008). Pattern selection in plants: coupling chemical dynamics to surface growth in three dimensions. *Ann. Bot.* **101**, 361–74.

Holloway, D. M., Harrison, L. G. and Armstrong, J. B. (1994). Computations of post-inductive dynamics in axolotl heart formation. *Dev. Dyn.* **200**, 242–56.

Holloway, D. M., Harrison, L. G. and Spirov, A. V. (2003). Noise in the segmentation gene network of *Drosophila*, with implications for mechanisms of body axis specification. *Proc. SPIE* **5110**, 180–91.

Holloway, D. M., Harrison, L. G., Kosman, D., Vanario-Alonso, C. E. and Spirov, A. V. (2006). Analysis of pattern precision shows that *Drosophila* segmentation develops substantial independence from gradients of maternal gene products. *Dev. Dyn.* **235**, 2949–60.

Holzinger, A., Callaham, D. A., Hepler, P. K. and Meindl, U. (1995). Free calcium in *Micrasterias*: local gradients are not detected in growing lobes. *Eur. J. Cell Biol.* **67**, 363–71.

Houchmandzadeh, B., Wieschaus, E. and Leibler, S. (2002). Establishment of developmental precision and proportions in the early *Drosophila* embryo. *Nature* **415**, 798–802.

Houchmandzadeh, B., Wieschaus, E. and Leibler, S. (2005). Precise domain specification in the developing *Drosophila* embryo. *Phys. Rev. E* **72**, 061920.

Howard, M. and Kruse, K. (2005). Cellular organization by self-organization: mechanisms and models for Min protein dynamics. *J. Cell Biol.* **168**, 533–6.

Howard, M. and Rein ten Wolde, P. (2005). Finding the center reliably: robust patterns of developmental gene expression. *Phys. Rev. Lett.* **95**, 208103.

Howard, M., Rutenberg, A.D. and de Vet, S. (2001). Dynamic compartmentalization of bacteria: accurate division in *E. coli. Phys. Rev. Lett.* **87**, 278102.

Humphrey, R.R. (1972). Genetic and experimental studies on a mutant (c) determining absence of heart action in embryos of the Mexican axolotl (*Ambystoma mexicanum*). *Dev. Biol.* **27**, 365–75.

Hunding, A. (1989). Turing patterns of the second kind simulated on supercomputers in three curvilinear coordinates and time. In *Cell to Cell Signalling: From Experiments to Theoretical Models*, ed. A. Goldbeter, pp. 229–36. London: Academic Press.

Hunding, A. (1993). Supercomputer simulation of Turing structures in *Drosophila* morphogenesis. In *Experimental and Theoretical Advances in Biological Pattern Formation*, eds. H.G. Othmer, P.K. Maini, and J.D. Murray, pp. 149–59. New York: Plenum Press.

Hunding, A. (2004). Microtubule dynamics may embody a stationary bipolarity forming mechanism related to the prokaryotic division site mechanism (pole-to-pole oscillations). *J. Biol. Phys.* **30**, 325–44.

Ingram, G.C., Goodrich, J., Wilkinson, M.D., Simon, R., Haughn, G.W. and Coen, E.S. (1995). Parallels between UNUSUAL FLORAL ORGANS and FIMBRIATA, genes controlling flower development in *Arabidopsis* and *Antirrhinum. Plant Cell* **7**, 1501–10.

Irish, V., Lehmann, R. and Akam, M. (1989). The *Drosophila* posterior-group gene *nanos* functions by repressing *hunchback* activity. *Nature* **338**, 646–8.

Isaacs, F.J, Blake, W.J. and Collins, J.J. (2005). Signal processing in single cells. *Science* **307**, 1886–88.

Isalan, M. (2009). Gene networks and liar paradoxes. *BioEssays* **31**, 1110–15.

Jacobson, A.G. (1960). Influences of ectoderm and endoderm on heart differentiation in the newt. *Dev. Biol.* **2**, 138–54.

Jacobson, A.G. and Duncan, J.T. (1968). Heart induction in salamanders. *J. Exp. Zool.* **167**, 79–103.

Jaeger, J. (2009). Modelling the *Drosophila* embryo. *Mol. BioSyst.* **5**, 1549–68.

Jaeger, J. and Goodwin, B.C. (2001). A cellular oscillator model for periodic pattern formation. *J. Theor. Biol.* **213**, 171–81.

Jaeger, J., Blagov, M., Kosman, D., *et al.* (2004a). Dynamical analysis of regulatory interactions in the gap gene system of *Drosophila melanogaster. Genetics* **167**, 1721–37.

Jaeger, J., Surkova, S., Blagov, M., *et al.* (2004b). Dynamic control of positional information in the early *Drosophila* embryo. *Nature* **430**, 368–71.

Jiang, T.X., Jung, H.S., Widelitz, R.B. and Chuong, C.M. (1999). Self-organization of periodic patterns by dissociated feather mesenchymal cells and the regulation of size, number and spacing of primordia. *Development* **126**, 4997–5009.

Jönsson, H., Heisler, M.G., Shapiro, B.E., Mjolsness, E. and Meyerowitz, E.M. (2006). An auxin-driven polarized transport model for phyllotaxis. *Proc. Nat. Acad. Sci. USA* **103**, 1633–8.

Jung, H.S., Francis-West, P.H., Widelitz, R.B., *et al.* (1998). Local inhibitory action of BMPs and their relationships with activators in feather formation: implications for periodic patterning. *Dev. Biol.* **196**, 11–23.

Kaandorp, J.A. (1994). *Fractal Modelling*. Berlin: Springer.

Kaern, M., Menzinger, M. and Hunding, A. (2000). Segmentation and somitogenesis derived from phase dynamics in growing oscillatory media. *J. Theor. Biol.* **207**, 473–93.

Kaern, M., Elston, T.C., Blake, W.J. and Collins, J.J. (2005). Stochasticity in gene expression: from theories to phenotypes. *Nature Rev. Genet.* **6**, 451–64.

Kaern, M., Menzinger, M., Satnoianu, R. and Hunding, A. (2001). Chemical waves in open flows of active media: their relevance to axial segmentation in biology. *Faraday Disc.* **120**, 295–312.

Käfer, J., Hogeweg, P. and Marée, A. F. M. (2006). Moving forward moving backward: directional sorting of chemotactic cells due to size and adhesion differences. *PLoS Comput. Biol.* **2**, e56.

Kaplan, D. R. (1992). The relationship of cells to organisms in plants: problem and implications of an organismal perspective. *Int. J. Plant Sci.* **153**, S28–S37.

Kauffman, S. A., Shymko, R. M. and Trabert, K. (1978). Control of sequential compartment formation in *Drosophila*. *Science* **199**, 259–70.

Kauzmann, W. (1957). *Quantum Chemistry*. New York: Academic Press.

Kent, L. J. and Knight, E. C., eds. (1969). *Selected Writings of E. T. A. Hoffmann*, 2 vols. Chicago and London: University of Chicago Press.

Kerszberg, M. and Wolpert, L. (2007). Specifying positional information in the embryo: looking beyond morphogens. *Cell* **130**, 205–9.

Kiermayer, O. (1964). Untersuchungen über die Morphogenese und Zellwandbildung bei *Micrasterias denticulata* Bréb. *Protoplasma* **59**, 382–420.

Kiermayer, O. (1967). Das Septum-Initialmuster von *Micrasterias denticulata* und seine Bildung. *Protoplasma* **64**, 481–4.

Kiermayer, O. (1968). Hemmung der Kern- und Chloroplasten-migration von *Micrasterias* durch Colchicin. *Naturwissenschaften* **55**, 299–300.

Kiermayer, O. (1970). Causal aspects of cytomorphogenesis in *Micrasterias*. *Ann. NY Acad. Sci.* **175**, 686–701.

Kiermayer, O. (1981). Cytoplasmic basis of morphogenesis in *Micrasterias*. In *Cytomorphogenesis in Plants*, ed. O. Kiermayer. Wien and New York: Springer.

Knight, J. (2003). Scientific literacy: clear as mud. *Nature* **423**, 376–8.

Ko, M. S. H. (1991). A stochastic model for gene induction. *J. Theor. Biol.* **153**, 181.

Ko, M. S. H. (1992). Induction mechanism of a single gene molecule: stochastic or deterministic? *BioEssays* **14**, 341–6.

Kondo, S. and Asai, R. (1995). A reaction–diffusion wave on the skin of the marine angelfish *Pomocanthus*. *Nature* **376**, 765–8.

Kondratiev, V. N. (1969). *The Theory of Kinetics*. New York: Elsevier.

Kosman, D., Reinitz, J. and Sharp, D. H. (1997). Automated assay of gene expression at cellular resolution. In *Proceedings of the 1998 Pacific Symposium on Biocomputing*, eds R. Altman, K. Dunker, L. Hunter and T. Klein, pp. 6–17. Singapore: World Scientific Press.

Kosman, D., Small, S. and Reinitz, J. (1998). Rapid preparation of a panel of polyclonal antibodies to *Drosophila* segmentation proteins. *Dev. Genes Evol.* **208**, 290–4.

Kuhn, T. S. (1962). *The Structure of Scientific Revolutions*. Chicago: University of Chicago Press.

Kuijt, J. (1967). On the structure and origin of the seedling of *Psittacanthus schiedeanus* (Loranthaceae). *Can. J. Bot.* **45**, 1497–506.

Kwiatkowska, D. and Dumais, J. (2003). Growth and morphogenesis at the vegetative shoot apex of *Anagallis arvensis* L. *J. Exp. Bot.* **54**, 1585–95.

Lacalli, T. C. (1973a). *Morphogenesis in Micrasterias*. PhD thesis, University of British Columbia, Canada.

Lacalli, T. C. (1973b). Cytokinesis in *Micrasterias rotata*. *Protoplasma* **78**, 433–42.

Lacalli, T. C. (1975). Morphogenesis in *Micrasterias*. I: Tip growth. II: Patterns of morphogenesis. *J. Embryol. Exp. Morphol.* **33**, 95–116, 117–27.

Lacalli, T. C. (1981). Dissipative structures and morphogenetic pattern in unicellular algae. *Phil. Trans. R. Soc. Lond. B* **294**, 547–88.

Lacalli, T.C. (1990). Modeling the *Drosophila* pair-rule pattern by reaction–diffusion: gap input and pattern control in a 4-morphogen system. *J. Theor. Biol.* **144**, 171–94.

Lacalli, T.C. and Harrison, L.G. (1978a). The regulatory capacity of Turing's model for morphogenesis, with application to slime moulds. *J. Theor. Biol.* **70**, 273–95.

Lacalli, T.C. and Harrison, L.G. (1978b). Development of ordered arrays of cell wall pores in Desmids: a nucleation model. *J. Theor. Biol.* **74**, 109–38.

Lacalli, T.C. and Harrison, L.G. (1991). From gradients to segments: models for pattern formation in early *Drosophila* embryogenesis. *Sem. Dev. Biol.* **2**, 107–17.

Lacalli, T.C., Wilkinson, D.A. and Harrison, L.G. (1988). Theoretical aspects of stripe formation in relation to *Drosophila* segmentation. *Development* **104**, 105–13.

Lander, A.D. (2007). Morpheus unbound: reimagining the morphogen gradient. *Cell* **128**, 245–56.

Lander, A.D., Nie, Q. and Wan, F.Y. (2002). Do morphogen gradients arise by diffusion? *Dev. Cell* **2**, 785–96.

Lander, A.D., Lo, W.-C., Nie, Q. and Wan, F.Y.M. (2009). The measure of success: constraints, objectives, and tradeoffs in morphogen-mediated patterning. *Cold Spring Harb. Perspect. Biol.* **1**, a002022.

Laskowski, M., Grieneisen, V.A., Hofhuis, H., *et al.* (2008). Root system architecture from coupling cell shape to auxin transport. *PLoS Biol.* **6**, e307.

Lengyel, I. and Epstein, I.R. (1991). Modeling of Turing structures in the chlorite–iodide–malonic acid–starch reaction system. *Science* **251**, 650–2.

Li, R. and Bowerman, B., eds. (2010). *Symmetry Breaking in Biology.* Cold Spring Harbor: Cold Spring Harbor Laboratory Press.

Lisman, J.E. (1985). A mechanism for memory storage insensitive to molecular turnover: a bistable autophosphorylating kinase. *Proc. Nat. Acad. Sci. USA* **82**, 3055–7.

Lopes, F.J.P., Vieira, F.M.C., Holloway, D.M., Bisch, P.M. and Spirov, A.V. (2008). Spatial bistability generates hunchback expression sharpness in the Drosophila embryo. *PLoS Comp. Biol.* **4**, e1000184.

Lott, S.E., Kreitman, M., Palsson, A., Alekseeva, E. and Ludwig, M.Z. (2007). Canalization of segmentation and its evolution in *Drosophila*. *Proc. Nat. Acad. Sci. USA* **104**, 10926–31.

Lucchetta, E.M., Vincent, M.E. and Ismagilov, R.F. (2008). A precise Bicoid gradient is nonessential during cycles 11–13 for precise patterning in the *Drosophila* blastoderm. *PLoS ONE* **3**, e3651.

Lucchetta, E.M., Lee, J.H., Fu, L.A., Patel, N.H. and Ismagilov, R.F. (2005). Dynamics of *Drosophila* embryonic patterning network perturbed in space and time using microfluidics. *Nature* **434**, 1134–8.

Lyons, M.J. and Harrison, L.G. (1991). A class of reaction–diffusion mechanisms which preferentially select striped patterns. *Chem. Phys. Lett.* **183**, 158–64.

Lyons, M.J. and Harrison, L.G. (1992a). Stripe selection: an intrinsic property of some pattern-forming models with nonlinear dynamics. *Dev. Dyn.* **195**, 201–15.

Lyons, M.J. and Harrison, L.G. (1992b). Non-linear analysis of models for biological pattern formation: application to ocular dominance stripes. In *Analysis and Modeling of Neural Systems II*, ed. F.H. Eeckman. Norwell, MD: Kluwer Academic.

Lyons, M.J., Harrison, L.G., Lakowski, B.C. and Lacalli, T.C. (1990). Reaction–diffusion modelling of biological pattern formation: application to the embryogenesis of *Drosophila melanogaster*. *Can. J. Phys.* **68**, 772–7.

Maini, P. K., Baker, R. E. and Chuong, C. M. (2006). The Turing model comes of molecular age. *Science* **314**, 1397–8.

Mandoli, D. F. (1998). Elaboration of body plan and phase change during development of *Acetabularia*: how is the complex architecture of a giant unicell built? *A. Rev. Plant Physiol. Plant Mol. Biol.* **49**, 173–98.

Manu, Surkova S., Spirov, A. V., Gursky, V. V., *et al.* (2009a). Canalization of gene expression in the *Drosophila* blastoderm by gap gene cross regulation. *PLoS Biology* **7**, e1000049.

Manu, Surkova S., Spirov, A. V., Gursky, V. V., *et al.* (2009b). Canalization of gene expression and domain shifts in the *Drosophila* blastoderm by dynamical attractors,*PLoS Comp. Biol.* **5**, e1000303.

Marée, A. F. M. and Hogeweg, P. (2001). How amoeboids self-organize into a fruiting body: multicellular coordination in *Dictyostelium discoideum*. *Proc. Nat. Acad. Sci. USA.* **98**, 3879–83.

Marée, A. F. M. and Hogeweg, P. (2002). Modelling *Dictyostelium discoideum* morphogenesis: the culmination. *Bull. Math. Biol.* **64**, 327–53.

Marée, S. (2000). *From Pattern Formation to Morphogenesis: Multicellular Coordination in Dictyostelium discoideum.* PhD thesis. University of Utrecht, Netherlands.

Martens, G., Humphrey, E. C., Harrison, L. G., *et al.* (2009). High-pressure freezing of spermiogenic nuclei supports a dynamic chromatin model for the histone-to-protamine transition. *J. Cell. Biochem.* **108**, 1399–409.

Matela, R. J. and Fletterick, R. J. (1980). Computer simulation of cellular self-sorting: a topological exchange model. *J. Theor. Biol.* **84**, 673–90.

McNally, J. G. and Cox, E. C. (1989). Spots and stripes: the patterning spectrum in the cellular slime mould *Polysphondylium pallidum*. *Development* **105**, 323–33.

Meindl, U. (1982). Local accumulations of membrane associated calcium according to cell pattern formation in *Micrasterias denticulata*, visualized by chlorotetracycline fluorescence. *Protoplasma* **110**, 143–6.

Meinhardt, H. (1982). *Models of Biological Pattern Formation.* London: Academic Press.

Meinhardt, H. (1984). Models of pattern formation and their application to plant development. In *Positional Controls in Plant Development*, ed. P. W. Barlow and D. J. Carr, pp. 1–32. Cambridge: Cambridge University Press.

Meinhardt, H. (1995). *The Algorithmic Beauty of Seashells.* 1st edn, Berlin, Heidelberg and New York: Springer-Verlag. (2nd edn 1998; 3rd edn, 2003).

Meinhardt, H. (2001). Organizer and axes formation as a self-organizing process. *Int. J. Dev. Biol.* **45**, 177–88.

Meinhardt, H. and de Boer, P. A. J. (2001). Pattern formation in *Esherichia coli*: a model for pole-to-pole oscillations of min proteins and the localization of the division site. *Proc. Nat. Acad. Sci. USA* **98**, 14202–7.

Mills, W. H. (1932). Stereochemistry and catalysis. *Chem. Ind.*, 750–9.

Milo, R., Shen-Orr, S., Itzkovitz, S., Kashtan, N., Chklovskii, D. and Alon, U. (2002). Network motifs: simple building blocks of complex networks. *Science* **298**, 824–7.

Mitchison, G. J. (1980). A model for vein formation in higher plants. *Proc. R. Soc. Lond. B* **207**, 79–109.

Mitchison, G. J. (1981). The polar transport of auxin and vein patterns in plants. *Phil. Trans. R. Soc. Lond. B* **295**, 461–71.

Mizutani, C. M., Nie, Q., Wan, F. Y. M., *et al.* (2005). Formation of the BMP activity gradient in the *Drosophila* embryo. *Dev. Cell* **8**, 915–24.

Moore, K. L. (1974). *The Developing Human: Clinically Oriented Embryology.* Philadelphia, London and Toronto: W. B. Saunders Co.

Moore, W. J. (1972). *Physical Chemistry*, 4th edn. Englewood Cliffs, N J: Prentice-Hall.

Muratov, C. B. and Shvartsman, S. Y. (2003). An asymptotic study of the inductive pattern formation mechanism in *Drosophila* egg development. *Physica D* **186**, 93–108.

Murray, J. D. (1981a). A pre-pattern formation mechanism for animal coat markings. *J. Theor. Biol.* **88**, 161–99.

Murray, J. D. (1981b). On pattern formation mechanisms for lepidopteran wing patterns and mammalian coat markings. *Phil. Trans. R. Soc. Lond. B* **295**, 473–96.

Murray, J. D. (1988). How the leopard gets its spots. *Sci. Am.* **258**, 80–7.

Murray, J. D. (1989). *Mathematical Biology*. Berlin: Springer-Verlag.

Nagata, W., Harrison, L. G. and Wehner, S. (2003). Reaction–diffusion models of growing plant tips: bifurcations on hemispheres. *Bull. Math. Biol.* **65**, 571–607.

Nägeli, C. (1847). *Die neuern Algensysteme*. Zürich, Switzerland: Schulthess.

Nagorcka, B. N. (1988). A pattern formation mechanism to control spatial organization in the embryo of *Drosophila melanogaster*. *J. Theor. Biol.* **132**, 277–306.

Nagorcka, B. N. and Mooney, J. R. (1982). The role of a reaction–diffusion system in the formation of hair fibres. *J. Theor. Biol.* **98**, 575–607.

Nagorcka, B. N. and Mooney, J. R. (1985). The role of a reaction–diffusion system in the initiation of primary hair follicles. *J. Theor. Biol.* **114**, 243–72.

Neville, A. A., Matthews, P. C. and Byrne, H. M. (2006). Interactions between pattern formation and domain growth. *Bull. Math. Biol.* **68**, 1975–2003.

Newman, S. A. (1988). Lineage and pattern in the developing vertebrate limb. *Trends Genet.* **4**, 329–32.

Newman, S. A. (1996). Sticky fingers: *Hox* genes and cell adhesion in vertebrate limb development. *BioEssays* **18**, 171–4.

Newman, S. A. and Frisch, H. L. (1979). Dynamics of skeletal pattern formation in developing chick limb. *Science* **205**, 662–8.

Nicolis, G. and Prigogine, I. (1977). *Self-Organization in Non-Equilibrium Systems*. New York: Wiley.

Nieuwkoop, P. D. (1969). The formation of the mesoderm in urodele amphibians: I. Induction by the endoderm. *Wilhelm Roux Arch. Entwicklungsmech. Org.* **162**, 341–73.

Nieuwkoop, P. D. (1973). The 'organisation center' of the amphibian embryo: its origin, spatial organisation and morphogenetic action. *Adv. Morphogenet.* **10**, 1–310.

Nieuwkoop, P. D. (1977). Origin and establishment of embryonic polar axes in amphibian development. *Curr. Top. Dev. Biol.* **11**, 115–32.

Nieuwkoop, P. D., Johnen, A. G. and Albers, B. (1985). *The Epigenetic Nature of Early Chordate Development*. Cambridge: Cambridge University Press.

Nishimura, N. J. and Mandoli, D. F. (1992). Vegetative growth of *Acetabularia acetabulum* (Chlorophyta): structural evidence for juvenile and adult phases in development. *J. Phycol.* **28**, 669–77.

Okabe-Oho, Y., Murakami, H., Oho, S. and Sasai, M. (2009). Stable, precise, and reproducible patterning of Bicoid and Hunchback molecules in the early *Drosophila* embryo. *PLoS Comp. Biol.* **5**: e1000486.

Orchard, J. J. (1996). *Reaction–diffusion Modelling of Somite Formation: Computed Dynamics and Bifurcation Analysis*. MSc thesis, University of British Columbia, Canada.

Oster, G. F., Murray, J. D. and Harris, A. K. (1983). Mechanical aspects of mesenchymal morphogenesis. *J. Embryol. Exp. Morphol.* **78**, 83–125.

Othmer, H. and Pate, E. (1980). Scale invariance in reaction–diffusion models of spatial pattern formation. *Proc. Nat. Acad. Sci. USA* **77**, 4180–4.

Ouyang, Q. and Swinney, H. L. (1991). Transition from a uniform state to hexagonal and striped Turing patterns. *Nature* **352**, 610–11.

Page, K. M., Maini, P. K., Monk, N. A. M. and Stern, C. D. (2001). A model of primitive streak initiation in the chick embryo, *J. Theor. Biol.* **208**, 419–38.

Palmeirim, I., Henrique, D., Ish-Horowicz, D. and Pourquié, O. (1997). Avian hairy gene expression identifies a molecular clock linked to vertebrate segmentation and somitogenesis. *Cell* **91**, 639–48.

Palsson, E. (2007). Modeling wave propagation, chemotaxis, cell adhesion and cell sorting: examples with *Dictyostelium*, using a 3-D cell-based model. In *Modeling Biology, Structures, Behaviors, Evolution, Vienna Series in Theoretical Biology*, ed. M. D. Laubichler and G. D. Muller, pp. 165–94. Cambridge, MA: MIT Press.

Palsson, E. and Cox, E. C. (1997). Selection for spiral waves in the social amoebae *Dictyostelium.Proc. Nat. Acad. Sci. USA* **94**, 13719–23.

Pankratz, M. J., Gaul, U., Hoch, M., *et al.* (1990). Overlapping gene activities generate pair-rule stripes and delimit the expression domains of homeotic genes along the longitudinal axis of the *Drosophila* blastoderm embryo. In *Genetics of Pattern Formation and Growth Control*, ed. A. P. Mahowald, pp. 17–29. New York: Wiley-Liss.

Papatsenko, D. (2009). Stripe formation in the early fly embryo: principles, models, and networks. *BioEssays* **31**, 1172–80.

Peck, A. L. (trans.) (1942). *Aristotle: Generation of Animals*. Cambridge, MA: Harvard University Press.

Pedraza, J. M. and van Oudenaarden, A. (2005). Noise propagation in gene networks. *Science* **307**, 1965–9.

Peel, A. (2004). The evolution of arthropod segmentation mechanisms. *Bioessays* **26**: 1108–16.

Peel, A., Chipman, A. D. and Akam, M. (2005). Arthropod segmentation: beyond the *Drosophila* paradigm. *Nat. Rev. Genet.* **6**, 905–16.

Phillips, H. M. and Steinberg, M. S. (1969). Equilibrium measurements of embryonic chick cell adhesiveness: I. Shape equilibrium in centrifugal fields. *Proc. Nat. Acad. Sci. USA* **64**, 121–7.

Pick, L. (1998). Segmentation: painting stripes from flies to vertebrates. *Dev. Genetics* **23**, 1–10.

Potts, R. B. (1952). Some generalized order–disorder transformations. *Proc. Camb. Phil. Soc.* **48**, 106–9.

Pourquié, O. (2003). The segmentation clock: converting embryonic time into spatial pattern. *Science* **301**, 328–30.

Poustelnikova, E., Pisarev, A., Blagov, M., Samsonova, M. and Reinitz, J. (2004). A database for management of gene expression data in situ. *Bioinformatics* **20**, 2212–21.

Prescott, G. W., Croasdale, H. T. and Vinyard, W. C. (1977). *A Synopsis of North American Desmids: II. Desmidiaceae: Placodermae, section 2*. Lincoln and London: University of Nebraska Press.

Prigogine, I. (1967). Dissipative structures in chemical systems. In *Fast Reactions and Primary Processes in Chemical Kinetics*, ed. S. Claesson, pp. 371–82. New York: Interscience.

Prigogine, I. and Lefever, R. (1968). Symmetry-breaking instabilities in dissipative systems: II. *J. Chem. Phys.* **48**, 1695–700.

Pueyo, J. I., Lanfear, R. and Couso, J. P. (2008). Ancestral Notch-mediated segmentation revealed in the cockroach *Periplaneta americana*. *Proc. Nat. Acad. Sci. USA* **105**, 16614–19.

Puiseux-Dao, S. (1962). Recherches biologiques et physiologiques sur quelques Dasycladacées, en particulier, le *Batophora oerstedii* et *l'Acetabularia mediterranea* Lam. *Rev. Gen. Bot.* **69**, 409–503.

Rao, C.V. and Arkin, A.P. (2001). Control motifs for intracellular regulatory networks. *Annu. Rev. Biomed. Eng.* **3**, 391–419.

Rao, C.V., Wolf, D.M. and Arkin, A.P. (2002). Control, exploitation and tolerance of intracellular noise. *Nature* **420**, 231–7.

Rashevsky, N. (1940). An approach to the mathematical biophysics of biological self-regulation and of cell polarity. *Bull. Math. Biophys.* **2**, 15–25.

Reddy, V.G. and Meyerowitz, E.M. (2005). Stem-cell homeostasis and growth dynamics can be uncoupled in the *Arabidopsis* shoot apex. *Science* **310**, 663–7.

Reeves, G.T., Muratov, C.B., Schüpbach, T. and Shvartsman, S.Y. (2007). Quantitative models of developmental pattern formation. *Dev. Cell* **11**, 289–300.

Reinhardt, D., Pesce, E.-R., Stieger, P., *et al.* (2003). Regulation of phyllotaxis by polar auxin transport. *Nature* **426**, 255–60.

Reinitz, J. and Sharp, D.H. (1995). Mechanism of formation of *eve* stripes. *Mech. Dev.* **49**, 133–58.

Reinitz, J., Mjolsness, E. and Sharp, D.H. (1995). Model for cooperative control of positional information in *Drosophila* by *bicoid* and *hunchback*. *J. Exp. Zool.* **271**, 47–56.

Rennich, S.C. and Green, P.B. (1997). The mathematics of plate bending. In *The Dynamics of Cell and Tissue Motion*, eds W. Alt, A. Deutsch and G. Dunn, pp. 251–3. Basel: Birkhauser Verlag.

Rida, P.C., Le Minh N. and Jiang, Y.J. (2004). A Notch feeling of somite segmentation and beyond. *Dev. Biol.* **265**, 2–22.

Rolland-Lagan, A.-G., Bangham, J.A. and Coen, E. (2003). Growth dynamics underlying petal shape and asymmetry. *Nature* **422**, 161–3.

Rosenfeld, N., Young, J.W., Alon, U., Swain, P.S. and Elowitz, M.B. (2005). Gene regulation at the single-cell level. *Science* **307**, 1962–5.

Salazar-Ciudad, I. and Jernvall, J. (2004). How different types of pattern formation mechanisms affect the evolution of form and development. *Evol. Dev.* **6**, 6–16.

Salazar-Ciudad, I., Jernvall, J. and Newman, S.A. (2003). Mechanisms of pattern formation in development and evolution. *Development* **130**, 2027–37.

Sater, A.K. and Jacobson, A.G. (1990). The restriction of the heart morphogenetic field in *Xenopus laevis*. *Dev. Biol.* **140**, 328–36.

Saunders, T. and Howard, M. (2009a). Morphogen profiles can be optimized to buffer against noise. *Phys. Rev. E* **80**, 041902.

Saunders, T. and Howard, M. (2009b). When it pays to rush: interpreting morphogen gradients prior to steady-state. *Phys. Biol.* **6**, 046020.

Serrano, N. and O'Farrell, P.H. (1997). Limb morphogenesis: connections between patterning and growth. *Curr. Biol.* **7**, R186–R195.

Shipman, P.D. and Newell, A.C. (2005). Polygonal planforms and phyllotaxis on plants. *J. Theor. Biol.* **236**, 154–97.

Sick, S., Reinker, S., Timmer, J. and Schlake, T. (2006). WNT and DKK determine hair follicle spacing through a reaction–diffusion mechanism. *Science* **314**, 1447–50.

Simon, R., Carpenter, R., Doyle, S., Coen, E. (1994). FIMBRIATA controls flower development by mediating between meristem and organ identity genes. *Cell* **78**, 99–107.

Slack, J.M.W. (1983). *From Egg to Embryo*. Cambridge: Cambridge University Press.

Smith, R.S., Guyomarc'h, S., Mandel, T., *et al.* (2006). A plausible model of phyllotaxis. *Proc. Nat. Acad. Sci, USA* **103**, 1301–6.

Smith, S.C. and Armstrong, J.B. (1990). Heart induction in wild-type and cardiac mutant axolotls (*Ambystoma mexicanum*). *J. Exp. Zool.* **254**, 48–54.

Smith, S.C. and Armstrong, J.B. (1991). Heart development in normal and cardiac-lethal mutant axolotls: a model for the control of vertebrate cardiogenesis. *Differentiation* **47**, 129–34.

Smith, S. C. and Armstrong, J. B. (1993). Reaction–diffusion control of heart development: evidence for activation and inhibition in precardiac mesoderm. *Dev. Biol.* **160**, 535–42.

Solms-Laubach, H. Graf zu. (1895). Monograph of the Acetabularieae. *Trans. Linn. Soc. Lond. 2 (Botany)* **5**, 1–39.

Solursh, M., Jensen, K. L., Zanetti, N. C., Linsenmayer, T. F. and Reiter, R. S. (1984). Extracellular matrix mediates epithelial effects on chondrogenesis in vitro. *Dev. Biol.* **105**, 451–7.

Spirov, A. V. and Holloway, D. M. (2003). Making the body plan: precision in the genetic hierarchy of *Drosophila* embryo segmentation. *In. Silico. Biol.* **3**, 89–100.

Spirov, A. V., Lopes, F. J. P. and Holloway, D. M. (2008). Molecular fluctuations and interpreting spatial gradients, applied to Hunchback pattern formation. *Dev. Biol.* **319**, 589.

Spirov, A. V., Fahmy, K., Schneider, M., *et al.* (2009). Formation of the *bicoid* morphogen gradient: an mRNA gradient dictates the protein gradient. *Development* **136**, 605–14.

Staehelin, A. and Giddings, T. H. (1982). Membrane-mediated control of cell wall microfibrillar order. In *Developmental Order: Its Origin and Regulation*, ed. S. Subtelny and P. B. Green, pp. 133–47. New York: Alan R. Liss.

Steer, M. W. (1988). The role of calcium in exocytosis and endocytosis in plant cells. *Physiologia Pl.* **72**, 213–20.

Steinberg, M. S. (1970). Does differential adhesion govern self-assembly processes in histogenesis? Equilibrium configurations and the emergence of a hierarchy among populations of embryonic cells. *J. Exp. Zool.* **173**, 395–434.

Streit, A., Berliner, A. J., Papanayotou, C., Sirulnik, A. and Stern, C. D. (2000). Initiation of neural induction by FGF signalling before gastrulation. *Nature* **406**, 74–8.

Surkova, S., Kosman, D., Kozlov, K., *et al.* (2008). Characterization of the *Drosophila* segment determination morphome. *Dev. Biol.* **313**, 844–62.

Swindale, N. V. (1980). A model for the formation of ocular dominance stripes. *Proc. R. Soc. Lond. B* **208**, 243–64.

Tardent, P. (1978). Coelenterata, Cnidaria. In *Morphogenese der Tiere*, pp. 199–302. Stuttgart: Gustav Fischer Verlag.

Thaller, C. and Eichele, G. (1987). Identification and spatial distribution of retinoids in the developing chick limb bud. *Nature* **327**, 625–8.

Thompson, D. W. (1917). *On Growth and Form*, 1st edn (2nd edn, 1942; abridged edn, ed. J. T. Bonner, 1961). Cambridge: Cambridge University Press.

Tickle, C., Summerbell, D. and Wolpert, L. (1975). Positional signalling and specification of digits in chick limb morphogenesis. *Nature* **254**, 199–202.

Tippit, D. H. and Pickett-Heaps, J. D. (1974). Experimental investigations into morphogenesis in *Micrasterias*. *Protoplasma* **81**, 271–96.

Tkacik, G., Gregor, T. and Bialek, W. (2008). The role of input noise in transcriptional regulation. *PLoS One* **3**, e2774.

Tostevin, F., ten Wolde, P. R. and Howard, M. (2007). Fundamental limits to position determination by concentration gradients. *PLoS Comp. Biol.* **3**, e78.

Townes, P. S. and Holtfreter, J. (1955). Directed movements and selective adhesion of embryonic amphibian cells. *J. Exp. Zool.* **128**, 53–120.

Turing, A. M. (1952). The chemical basis of morphogenesis. *Phil. Trans. R. Soc. Lond. B* **237**, 37–72.

Tyson, J. J. and Kauffman, S. A. (1975). Control of mitosis by a continuous biochemical oscillation. *J. Math. Biol.* **1**, 289–310.

Tyson, J. J. and Light, J. C. (1973). Properties of two-component bimolecular and trimolecular chemical reaction systems. *J. Chem. Phys.* **59**, 4164–73.

Ueda, K. and Noguchi, T. (1988). Microfilament bundles of F-actin and cytomorpho-genesis in the green alga *Micrasterias crux-melitensis. Eur. J. Cell Biol.* **46**, 61–7.

Ueda, K. and Yoshioka, S. (1976). Cell wall development of *Micrasterias americana*, especially in isotonic and hypertonic solutions. *J. Cell Sci.* **21**, 617–31.

Umulis, D. M., Serpe, M., O'Connor, M. B. and Othmer, H. G. (2006). Robust, bistable patterning of the dorsal surface of the *Drosophila* embryo. *Proc. Nat. Acad. Sci. USA* **103**, 11613–18.

Umulis, D. M., Shimmi, O., O'Connor, M. B. and Othmer, H. (2010). Organism-scale modeling of early *Drosophila* pattering via bone morphogenetic proteins. *Dev. Cell* **7**, 1–15.

Virag, A. (1999). *Discovering Genes Involved in Branching Decisions in Neurospora crassa.* PhD thesis, University of British Columbia, Canada.

Virag, A. and Griffiths, A. J. F. (2004). A mutation in the *Neurospora crassa* actin gene results in multiple defects in tip growth and branching. *Fungal genet. biol.* **41**, 213–25.

Vöchting, H. (1877). Ueber Theilbarkeit im Pflanzenreich und die Wirkung innerer un äusserer Krafte auf Organbildung an Pflanzentheilen. *Pflüger's Arch.* **15**, 153–90.

Vöchting, H. (1878). *Ueber Organbildung im Pflanzenreich*, Vol. **1**. Bonn, Germany: Max Cohen & Sohn.

von Aderkas, P. (2002). In vitro phenotypic variation in larch cotyledon number. *Int. J. Plant Sci.* **163**, 301–7.

Waddington, C. H. (1956). *Principles of Embryology.* London: Allen & Unwin.

Wald, G. (1957). The origin of optical activity in nature. *Ann. N. Y. Acad. Sci.* **69**, 352–67.

Walker, D. C., ed. (1979). *Origins of Optical Activity in Nature.* Amsterdam: Elsevier.

Waller, M. D. (1961). *Chladni Figures: A Study in Symmetry.* London: G. Bell & Sons.

Wang, Y. C. and Ferguson, E. L. (2005). Spatial bistability of Dpp-receptor inter-actions during *Drosophila* dorsal–ventral patterning. *Nature* **434**, 229–34.

Wässle, H. and Riemann, H. J. (1978). The mosaic of nerve cells in the mammalian retina. *Proc. R. Soc. Lond. B* **200**, 441–61.

Weil, T. T., Forrest, K. M. and Gavis, E. R. (2006). Localization of *bicoid* mRNA in late oocytes is maintained by continual active transport. *Dev. Cell* **11**, 251–62.

Weil, T. T., Parton, R., Davis, I. and Gavis, E. R. (2008). Changes in *bicoid* mRNA anchoring highlight conserved mechanisms during the oocyte-to-embryo transition. *Curr. Biol.* **18**, 1055–61.

Werz, G. (1965). Determination and realization of morphogenesis in *Acetabularia. Brookhaven Symp. Biol.* **18**, 185–203.

West, W. and West, G. S. (1905). *A Monograph of the British Desmidiaceae*, Vol. **2**. London: Adlard & Son, for the Ray Society.

Wigglesworth, V. B. (1940). Local and general factors in the development of 'pattern' in *Rhodnius prolixus* (Hemiptera). *J. Exp. Biol.* **17**, 180–200.

Wolff, C., Sommer, R., Schroeder, R., Glaser, G. and Tautz, D. (1995). Conserved and divergent expression aspects of the *Drosophila* segmentation gene hunch-back in the short germ band embryo of the flour beetle *Tribolium. Development* **121**, 4227–36.

Wolpert, L. (1970). Positional information and pattern formation. In *Towards a Theoretical Biology*, ed. C. H. Waddington, Vol. **3**, pp. 198–230. Edinburgh: Edinburgh University Press.

Wolpert, L. (1971). Positional information and pattern formation. In *Current Topics in Developmental Biology*, Vol. **6**, pp. 183–224. New York: Academic Press.

Wolpert, L. (1975). Control processes in development: pattern formation in chick limb morphogenesis. *Ann. Biomed. Eng.* **3**, 401–5.

Wolpert, L. (1981). Positional information and pattern formation. *Phil. Trans. R. Soc. Lond. B* **295**, 441–50.

Wolpert, L. (1990). The embryonic position. *The New Biologist* **2**, 1075–8.

Wolpert, L. (1992). Gastrulation and the evolution of development. *Development* **116**, 7–13.

Wolpert L. (2002). *Principles of development.* 2nd edn. London: Oxford University Press.

Wortis, M., Seifert, U., Berndl, K., *et al.* (1993). Curvature-controlled shapes of lipid-bilayer vesicles: budding, vesiculation and other phase transitions. In *Dynamical Phenomena at Interfaces, Surfaces and Membranes*, ed. D. Beysens, N. Boccara and G. Forgacs, pp. 221–36. Commack, N Y: Nova Science Publishers.

Wu, Y.F., Myasnikova, E. and Reinitz, J. (2007). Master equation simulation analysis of immunostained Bicoid morphogen gradient. *BMC Sys. Biol.* **1**, 52.

Zhabotinskii, A.M. (1964). Periodical oxidation of malonic acid in solution (a study of the Belousov reaction kinetics). *Biofizika* **9**, 306–11.